Charles Robert Richet, Edward Payson Fowler

Physiology and Histology of the Cerebral Convolutions

Charles Robert Richet, Edward Payson Fowler

Physiology and Histology of the Cerebral Convolutions

ISBN/EAN: 9783337373207

Printed in Europe, USA, Canada, Australia, Japan

Cover: Foto ©berggeist007 / pixelio.de

More available books at **www.hansebooks.com**

PHYSIOLOGY AND HISTOLOGY

OF THE

CEREBRAL CONVOLUTIONS.

ALSO,

POISONS OF THE INTELLECT,

BY

CHAS. RICHET, A.M., M.D., PH.D.,

(FORMER INTERN OF THE HOSPITAL OF PARIS.)

TRANSLATED BY

EDWARD P. FOWLER, M.D.

NEW YORK:

WM. WOOD & CO., 27 GREAT JONES ST.

1879.

STEAM PRESS OF
H. O. A. Industrial School,
76th St., near Third Ave.

OTHER WORKS BY THE SAME AUTHOR.

1. Recherches expérimentales et cliniques sur la sensibilité, in 8vo (Masson), 1877.

2. Les poisons de l'intelligence, in 12mo (P. Ollendorf),
1877.

3. Le somnambulism provoqué (*Jour. de l'anatomie et de
la physiologie*), 1877.

4. Etude sur la douleur (*Revue Philosophique*), 1877.

5. Essai sur les causes de dégoût (*Revue des Deux Mondes*),
1877.

6. Du suc gastrique chez l'homme et les animaux, ses
propriétés chimiques et physiologiques, in 8vo (Germer,
Baillière et Cie.), 1878.

AUTHOR'S PREFACE

SHOULD my work appear to its readers in some degree incomplete, as it certainly must, I will solicit indulgence upon the following grounds:

1st. That discoveries in cerebral physiology succeed each other with exceptional rapidity, and any work upon this subject, after a few months of existence, of course fails to include an important mass of facts which each new day develops.

2d. To fairly understand this department in medical science, and to be able to explain it intelligently, exacts a familiarity with a greater number of sciences than does almost any other subject.

First of all, it is, as a matter of course, requisite to be a PHYSIOLOGIST: the most important results are derived from vivisections, but it must be learned (no small task) both how to make and to interpret them.

There must also be a knowledge of SURGICAL PATHOLOGY in order to discriminate between cerebral commotions resulting from surgical processes (trephining, etc.), and those depending upon other (physiological) causes.

Joined with this, proficiency in MEDICAL PATHOLOGY is imperative, in order to recognize those pathological conditions (cerebral atrophy, general paralysis, cortical paralysis, aphasia, etc.) which are so inseparable from this study.

There must be, too, a degree of acquaintance with the science of PHYSICS, as exemplified in understanding electric excitations, their number, frequency, diffusion, polarization, etc.

As for ANATOMY, a thorough knowledge of that which concerns the human subject is but the introduction ; comparative anatomy, histology, embryology, and physiology are of still vaster importance, and for this purpose we are forced to the study of another science, ZOÖLOGY. Having surveyed the wide field of zoölogy and returned to the culminating object of our study—man—necessary comparisons between human beings and races are impossible without the aid of another science, that of ANTHROPOLOGY.

Again, in cerebral physiology the science of PSYCHOLOGY is especially requisite, for no one who has not deeply reflected upon the processes of intellection is capable of producing a good cerebral physiology ; the very essence of the subject would be to him a closed volume. To be sure, this science, notwithstanding the many admirable works written, and the labors which many profound thinkers have bestowed upon it, is as yet in its rough outline. *The laws of human thought*, what is more mysterious! They lie at the very foundation of our subject. The movement of the heart is its physiological function ; thought is the physiological function of the brain. Now the movement of the heart, though relatively easy to see and study, has required centuries of gropings and errors to become understood, and is there not much greater reason to anticipate like obstacles to a complete understanding of the action of thought, a subject so difficult to examine and analyze?

The list might be indefinitely multiplied, and in all these sciences, each one of which is a life-long study, who can hope to be so perfected as not to be justly exposed to criticism?

I have been and perhaps will again be charged by my readers with a lack of positiveness, in other words of being SKEPTICAL. But the very accusation seems to me eulogistic, for in science there is nothing more baneful than to treat hypotheses as certainties. On the contrary, when serious criticism has revealed the defects and feebleness of an experiment, a real service has been rendered, for it may incite to new experiments and unequivocal conclusions.

Inductions from probabilities or ill-demonstrated experiments are unreliable, and intelligent skepticism is more valuable to the advance of science than unbridled enthusiasm.

In connection with our subject, I would refer to one danger which should be guarded against and which has been somewhat overlooked. That is, recent labors have been accepted too much to the exclusion or neglect of those further in the past.

It is an unfortunate tendency and one that results in injustice. For example: in the physiological history of the convolutions, some of the finest discoveries were made by Flourens. There are few experiments as interesting as that in which the pigeon, deprived of its cerebral lobes, *sits plunged into a profound sleep of everlasting unconsciousness.*

I most fully recognize that recent investigators, Fritsche and Hitzig, Ferrier, Charcot, and others have made magnificent discoveries: still it is Flourens who stands in the front rank, and it need not be considered that the science of cerebral physiology dates from 1872.

———

I am gratified that Dr. Fowler, who has already so ably translated Charcot's work upon LOCALIZATION IN DISEASES OF THE BRAIN, has deemed my book worthy the same consideration. He has my full appreciation of the compliment and of the conscientious and scholarly manner in which his labor has been performed.

Lastly, I do not feel that, in addressing itself to the American profession, my volume is going among strangers, for in science there is but one land and one people.

CH. RICHET.

PARIS, April 29, 1879.

TRANSLATOR'S PREFACE.

EVERY student in medicine knows that an intelligent recognition of pathology, and a judicious, rational management of disease, are in exact ratio to a precise knowledge of normal anatomy and physiology.

For this reason, Richet's PHYSIOLOGY AND HISTOLOGY OF THE CEREBRAL CONVOLUTIONS seems a natural complement to Charcot's "LOCALIZATION IN DISEASES OF THE BRAIN."

In the anatomical portion, Richet includes the latest researches.

As regards physiology, he indicates those points which are settled beyond dispute; but where the matter is still under discussion, the arguments upon both sides are faithfully given, and the author conscientiously abstains from lending bias by way of mere personal opinion or theory; he is preëminently a dealer in facts.

The author suggested that an abridged form of a little monograph which he has written upon "POISONS OF THE INTELLECT" would add somewhat to the interest of the physiological part of the book, and in compliance therewith, I have made the addition, which my friend Dr. John C. Minor has very kindly and ably translated and arranged. It was originally written as a popular article for non-professional readers, and hence the author has presented the subject as a study in psychology rather than in the more exact sciences of anatomy and physiology.

In order to bring the subject within the required limits, it has been necessary to condense as much as the preservation of the original plan of arrangement would permit, and at the same time to eliminate everything that was not essential to the subject. Hence, perhaps the most interesting portion of the work, that which dealt particularly with illustrative cases and extended descriptions, was unwillingly sacrificed in order to present the work in its abridged form.

NEW YORK, June, 1879.

INDEX OF ILLUSTRATIONS.

TABLE OF CONTENTS.

CEREBRAL CONVOLUTIONS.

FIRST PART.

STRUCTURE OF THE CONVOLUTIONS.

I. HISTORY.

ALTHOUGH the arrangement and the morphology of the cerebral convolutions are now tolerably well known, their structure has not been described with corresponding exactitude, and our latest knowledge upon the subject is still defective. We will first examine the opinions of the ancient anatomists.

Hippocrates[1] compares the brain to a gland : *caput quoque ipsum glandulas habet; nam cerebrum glandulæ simile.*

According to Malpighi[2] and Vieussens[3] also, the cerebral cortex was a gland.

With some reservations, Malpighi compares the cerebral glands to the hepatic lobule,—*Non improbabile interim putans iisdem etiam acinis* (of the liver) *has cerebri glandulas conglobari posse.*[4]

Indeed, Malpighi supposed the brain to be composed of fibres and glands —*Corticales glandulæ tortuose poœatœ exteriores cerebri gyros componunt, et exorientibus inde medullaribus fibris, seu vasculis, appenduntur, ita ut ubicunque per transversum secantur gyri, determinata et firma semper glandularum congerie medullæ affundatur.*

---

[1] Cited by Longet, Anat. et. physiol. du système nerveux, I., p. 160.

[2] De cerebri cortice dissertatio, in Bibl. anat. de Manget, t. ii., p. 82. Genève, 1699.

[3] Institut. de méd., t. iii., p. 109, 2d edition.

[4] Loc. cit., p. 323.

I do not know that one can be as positive as Luys,[1] and affirm that Malpighi discovered the brain-cells; yet it seems to me very probable. In another place he says: *glandularum exiguitas aciem microscopii subterfugit;* probably signifying that he expected to see the lobules, but has only seen simple, very small, non-agglomerated cells (?).

However it may be, the ideas of Malpighi were soon forgotten. In 1698, Ruysch,[2] from the results of his marvellous arterial injections, considered the cerebral cortex as a vascular network. That opinion, advanced some time before by Leuwenhoeck, was generally adopted by anatomists. Boerhave,[3] who at first had admitted the opinion of Malpighi, finally adopted that of Ruysch.

The theory of Ruysch rested upon an exact fact: the extreme vascularity of the cerebral cortex. At first no objection was offered. At the time of Haller[4] (1766) the opinion was classic. It was admitted that there were arteries which could be injected, and others which were invisible and which could not be injected—all being plunged into a very fine cellular woof. There were distinguished in the brain a white portion (medulla) and a gray portion (cinereus cortex) made up by a mesh of arteries.

The first anatomist who gave a good idea of the cerebral structure was probably Vicq d'Azyr.[5] He pointed

---

[1] Le cerveau et ses fonctions, p. 15.

[2] Thesaurus anatom., vi., n. 73, thes. iv., n. 78, p. 78 ; p. 81 (citation of Haller, t. iv., epist. xi., p. 24).

[3] Cited by Haller, elementa phys., iv., p. 25.   [4] Loc. cit., p. 27.

[5] Vicq d'Azyr thus expresses himself: "In preparing the centrum ovale of Vieussens, if the form of the posterior cerebral convolutions which lie upon the tentorium cerebelli be examined, there generally will be noticed several that are remarkable on account of a white line which longitudinally divides the cortex, following all its contours, and which give to that portion of the cortex the appearance of a striped ribbon. I have found that arrangement in no other region of the brain." That remarkable observation of Vicq d'Azyr, though not overlooked, was not utilized, and it is but recently that Broca was the first to make apparent its importance.—*Upon the special structure of the inferior convolutions of the occipital lobe ; constant presence of the striped ribbon of Vicq d'Azyr.—Bull. de la soc. d'anthropologie,* 1861, *t. ii., p.* 313.

out that the gray cortex was really composed of three layers: a white layer between two gray layers, giving to the gray substance the appearance of a striped ribbon.

The fact was generally accepted, being exact and indisputable. Gennari,[1] however, admits the existence of a peculiar yellow substance between the central white substance and the cortex. Subsequent observations have shown that this yellow layer is in reality the deeper portion of the cortex.

In 1840 appeared the standard work of Baillarger[2]; that eminent anatomist announced that the gray portion of each convolution was composed of six layers, alternately gray and white.

To demonstrate his proposition, Baillarger examined the convolutions both by transmitted and direct light. When rendered transparent, the gray layers appeared clear, the white opaque. Baillarger showed that in young infants this appearance was very marked, and could be found even in the fœtus of four or five months.

It was about this time that the microscope was brought into use, and with that the minute structure of the convolutions could be investigated. The discovery of nerve-cells by Ehrenberg, Valentin, Purkinje (1835 to 1840) demonstrated the presence of nerve-cells very variable in form in the cerebral cortex.

Whatever may be the interest belonging to the revolutions from that time-to the present concerning the subject of nerve-tissue, I cannot stop to consider it.

Respecting the convolutions, Meynert is certainly the first who has given an exact description accompanied with good illustrations.

At the same period, Luys, in his grand work upon the nervous system, established the connection of the cortical and ganglionic systems of the brain. After Luys and Mey-

---

[1] Cited by Longet, loc. cit., t. i., p. 608.
[2] Recherches sur la structure de la couche corticale des convolutions du cer veau.—Mém. de l'Acad. de méd., t. viii., 1840.

nert ought to be cited the work of Betz, who first described certain cells in the motor-centres of the cortex. Also the staff of the Lunatic Assylum of London has recently published a series of valuable works upon the structure of the convolutions.

Reviewing this rapid history we find:

1st. That probably Malpighi discovered the cells and nerve-fibres of the cerebral cortex.

2d. That Vicq d'Azyr and Baillarger have well described the structure of the convolutions as seen by the naked eye.

3d. That Meynert, Luys, and Betz have made known the microscopic structure.

Sec. 2. GENERAL ARRANGEMENT OF THE CONVOLUTIONS.

If a human brain, divested of its membranes, be examined, it will be observed that the surface is of an ash-gray, covered with a multitude of furrows variable in dimensions, irregular in appearance, and which limit the projectio ns, the elevations of which are the cerebral convolutions.[1]

Thus each convolution forms an oblong mass, with blunt and rounded angles so confounded at each end with other convolutions that the precise point of commencement and ending can only be schematically designated. It rests against one or more of the neighboring convolutions, and the intervening furrow, empty when the pia mater and arachnoid are removed, is, when the membranes are in place, filled with vessels. The joining faces of the convolu-

---

[1] Gratiolet proposed to replace this unwieldly and inexact term (as applied to the lower mammifera at least) by the word fold (plis), but usage has not sanctioned the term. Perhaps, as Broca suggests, they might be called volutions.

tions are filled out, and they so join each other that the intermediate furrow is divided into two parts, the superior one for the veins, the inferior for the arteries. The entire gray cortex, here rising into a convolution, there sinking into a furrow, may be looked upon as a continuation of a single layer folded upon itself. This was Gall's idea, an ingenious one, and well calculated to express in schematic and easily-to-be-remembered manner the construction of the exterior cerebral covering.

The gray and white substance of the convolutions form a cortex sometimes called convolutionary, which should, according to Burdach and Broca,[1] be called palleum or manteau (mantle).

The remainder of the cerebral hemisphere forms the body of the brain.

The convolutions are separated by furrows of various depths, and according to their depths and their morphology they are differently named, fissures, furrows, creases.[2]

It should be especially observed that none of these fissures or furrows serve to completely separate adjoining convolutions, the continuity of which is never interrupted. To employ a common comparison, the cerebral convolutions represent a chain of mountains in which no valley traverses in such manner as to isolate the mountain peaks and plateaux. Nowhere upon the brain do two transverse, parallel furrows intersect two longitudinal parallel furrows. In short, all the furrows are incomplete—insufficient, as it were—as though they were unable to attain to the extremity of the convolution with which they commenced. Perhaps this may be of some importance in a functional point of view.

Broca thus defines the term convolution :[3]—" The word,

---

[1] Revue d'anthropologie, 1878, p. 197.

[2] See article, Circonvolutions (Pozzi), Dict. encyclop. des sciences méd., p. 342.

[3] Anatomie comparée des circonvolutions cérébrales.—Revue d'anthropologie, 1878, p. 391.

convolution, which has heretofore been employed to designate some portion of the folded surface of the brain, has now come to have an accepted meaning : it is applied to the subdivisions of lobes, and if in some cases a lobe can be composed of a single convolution, on the other hand no convolution can exceed the limits of its lobe, even though it may continue more or less directly with a convolution of a joining lobe."

As fissures differ in importance, so also do convolutions. To slightly marked furrows and to creases belong convolutions which are barely outlined, and which, according to Pozzi, may be called folds. But however greatly the furrows and convolutions may vary in size, the structure is always the same, and consequently a general description applies to the entire mass.

In certain parts of the brain the cortex folds upon itself, and by invagination forms an inverted convolution (gyrus Hippocampi), but here, too, the structure and arrangement are the same.

A point which should be well observed, but which we cannot properly dwell upon here, is that the convolutions, in most individuals at least, are not symmetrical.

SEC. 3. ORGANIZED ELEMENTS OF THE CONVOLUTIONS.

As observed by the anatomists of the last century, each convolution is composed of an external gray layer which exactly fits to the central white part : both receive vessels. We have, therefore, to study :—

1st. The gray layer, the cortex of the convolution (cortex cerebri).

2d. The white layer, which forms the axis.

3d. The vessels distributed to each layer.

The gray cortex probably is not a homogeneous layer; Baillarger[1] maintains that it is composed of six layers, being arranged from without towards the centre as follows:—

1. White layer.
2. Gray layer.
3. White layer.
4. Gray layer.
5. White layer.
6. Gray layer.

The four internal layers are often confounded in a single yellowish-red layer, by some authors described as a special layer.[2] Meynert holds to five layers.[3]

Mathias Duval[4] admits either five or six layers according as one counts the fifth and sixth as one or two.

Lewis[5] admits, in the human subject, five layers.[6]

All these layers of gray substance contain four varieties of organized elements: pyramidal, giant and fusiform cells, and granules (myélocytes).

1st. PYRAMIDAL CELLS.—These are the most numerous and are generally small and more difficult to discover than in the corresponding layers of the cerebellum. They are provided with several slender, pale, ramified prolongations of which three kinds can be distinguished.[7]

a. The pointed extremity of the cell, directed towards the periphery, continues by a fine thread which bends backwards[*] (pyramidal prolongation of Meynert).

---

[1] Loc. cit. et Comptes rendus de l'Ac. des sciences.
[2] See Kölliker, Elém. d'histol., French trans., p. 337.
[3] Journ. de l'anat., 1874, t. x., p. 100.—Pouchet, Traité d'hist., p. 307.
[4] Art. Nerveux (système) du Dict. de méd. et de chirurgie pratiques, p. 480.
[5] On the Comparative Structure of the Cortex Cerebri; Brain. 1878, p. 82.
[6] In Henle's Handbuch der Nervenlehre, 1871, p. 277, will be found a complete resumé of opinions expressed respecting the number of layers contained in the gray cortex.
[7] Pouchet and Tourneux: Traité d'histol., p. 307.—Meynert: Stricker's Handbuch, I., p. 708.
[*] This opinion, contrary to that of Bützke (Archiv. für Psychiatrie, 1872, t. iii., p. 300, has been admitted by Arndt (Arch. für microscop. Anat., 1874).

b. Laterally the cell gives from each side prolongations which are either oblique or perpendicular to its axis.

c. At the base of the cell is a prolongation analogous to the prolongation of Deiters (basal prolongation of Meynert).

These cells are generally in form of pyramids or triangles, the bases of which face the white substance, the points being directed towards the periphery.

These two cellular prolongations have been very well represented in their normal state by Mierzcjewski in two figures, which we here reproduce.

In the human subject it is difficult to see the basal prolongation, but that is no reason, as Arndt[1] has vainly endeavored, for putting in question its existence.

Naturally it is very difficult to follow to the terminations of the prolongations, and perhaps only a moderate degree of confidence should be given to the investigations of Golgi.[2] Golgi believed to have seen the basal prolongation of the pyramidal cells passing backwards, and, after a short distance, dividing and giving lateral branches, which, bending back again, ran to the periphery of the brain. The other prolongations are in connection with the conjunctive or granular cells. The anastomoses of the various parts were first illustrated by Luys, then by Besser[3] and Arndt.

According to Koschewnikoff, the basal prolongation may be followed into the white substance, where it is surrounded by myeline, and becomes the axis-cylinder of a nerve.[4]

Boll[5] considers the pyramidal form of the cells the result of the preparation: treated with osmic acid, which surprises them as it were living, they are circular.

---

[1] Arndt : Studien über die Architectonik der Grosshirnrinde, Max Schulze's Archiv, 1874.

[2] Sulla struttura della sostansa grigia del cervello. Communic. preventiva. Gaz. méd. ital. Lomb., ser. 6, t. vi.

[3] Eine Anastomose für die Centralen Ganglien-Zellen. Virchow's Archiv, Bd. 36.   [5] Arch. für Psych., 1873.

[4] Koschewnikoff : Arch. de Schultze, 1869, p. 332 et 375.

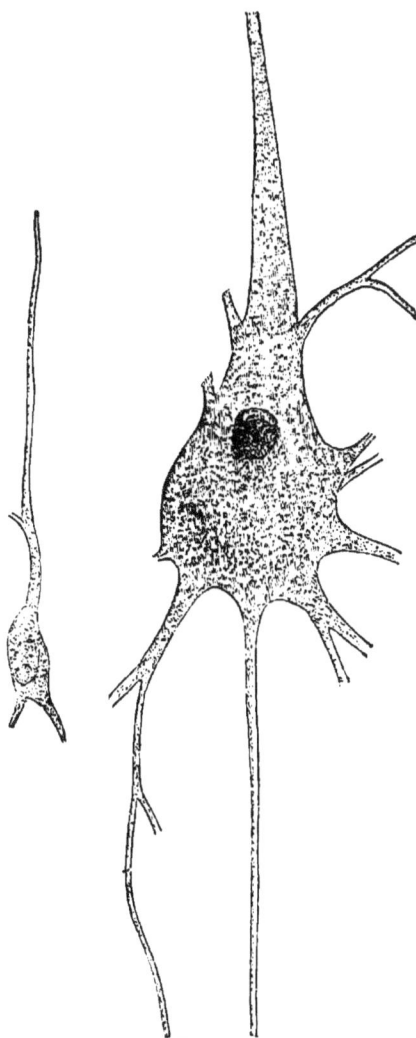

FIG. 1 (After Mierzejewski).—Pyramidal cell of the middle occipital lobe (solitary cell of Meynert).

FIG. 2.—Giant cell, common to the paracentral lobe. At the lower part, the basal prolongation ; at the upper part, the pyramidal prolongation.

They are often supplied with pigmentary granulations.

According to Luys,[1] they have, in fresh specimens, an amber-yellow color, and are provided with a brilliant nucleus and a nucleolus.

It seems possible to penetrate still further into the intimate structure of the cell. Without discussing this question, histologically so interesting, we would remark that the researches of Harless, Wagner, Flemming, Stark, and others indicate that the nerve-cell is a very complicated structure. The study of the minute structure of the nerve-cells, however, has generally been pursued in the spinal cord of the ox (anterior cornua), few observations having been made of the cells in the reticulum of the cerebral cortex.

Luys, however, has well described these cells, and below is the plate which he gives (Fig. 3).

That learned anatomist considers the cellular body to possess a truly reticulated structure. The reticulum is made up of fibrillæ, which, arranged like the trellis of a willow-basket, converge towards the nucleus of the cell.

Some of the pyramidal cells are small, 10 mm.; others average 22 mm. According to Luys, who first described them, their number approximates about 110 to the square millimetre—a considerable number, considering the surface of the brain, as compared to the small number of the medullary cells. Thus there are very many more cells in the brain than in the spinal cord.

It appears that the protoplasmic prolongations of all the cells anastomose in such way as to form a fine nerve-network, analogous to that found in the spinal cord (Gerlach and Boll.)

Bützke[2] considers a distinctive characteristic of the nerve-cells of the cortex cerebri to be the longitudinal striation of the cellular body, as well as of its prolongations; the same as transverse striations characterize muscular fibre.

[1] Le cervau, 1876, p. 14.   [2] Loc. cit. p. 589.

2d. GIANT-CELLS.—There is another variety of cells found in certain regions, which are remarkable on account of their size. They are faithfully described by Betz,[1] who calls them giant-cells.[2] They attain a diameter of 50 mm. Mierzejewski says that they always contain a yellowish-brown pigment.[3] It is generally believed that they are very

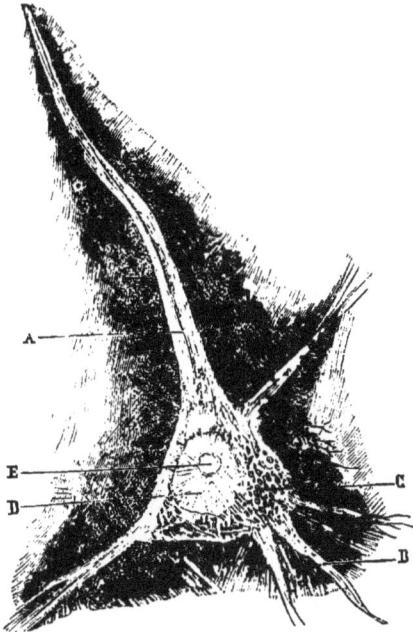

FIG. 3.

FIG. 3 (after Luys).—A cortical cell of the deep layers—about 800 diameters. The cell is divided in its long axix, and the interior texture can be seen.

A represents the superior prolongation, coming from the body of the cell. B, lateral and posterior prolongations. C, spongy, areolar substance, in which is found the cellular stroma. D, the cell seems to have the same thickness as the stroma; it sometimes has a radiated appearance. E, the brilliant nucleolus is also decomposable into secondary fibres.

[1] Anat. Nachw. zweier Gehirn-Centra.   Centralblatt, 1874, Nos. 37 and 38.
[2] Mierzejewski (Arch. de physiol., 1875, p. 226) observed them one year prior to Betz perhaps, but the determining of their topography undoubtedly belongs to Betz.   [3] Loc. cit., p. 228.

analogous to the pyramidal cells, from which they differ only in size.

3d. MYELOCYTES.—The myelocytes, or granules, are especially abundant in the lower layers of the cortex. They should not be regarded as cells, but as cellular nuclei, the cellular body which surrounds them being very reduced, and requiring special preparation in order to be seen. There is a lack of accord concerning the nature of these elements, though present opinion seems inclined to view them as embryonic nerve-cells.

Bützke holds that between the granules and the true conjunctive cells there is an entire series of intermediate bodies. Perhaps observation has here been too much adapted to the support of a theory.

It is possible that there may be in the gray cortex cerebri every transition between these cells (of which the nuclei are the granules) and the nerve-cells proper. Perhaps we should consider as the element of transition from the granule to the nerve-cell, these irregular, globular, non-pyramidal cells described by Meynert, and apparently resembling the elements of the granular layer of the retina.

All this is still obscure.

4th. FUSIFORM CELLS.—This variety has been described by Mernert, and prior to him by Berlin ;[1] they are fusiform (spindelförmig), generally bipolar, one prolongation being directed towards the periphery, the other towards the centre. According to Lewis, they are found in every part of the encephalon, and always in the most internal layer of the cortex, which they serve to characterize. They are very abundant in the claustrum (avant-mur), as we will see further on.

Between the fusiform, pyramidal, and giant-cells there exists a fine, granular, amorphous substance more or less abundant, a description of which has been given by Robin.

---

[1] Berlin : Beiträge zur Structurlehre der Gross-Gehirnwindungen. Erlangen, 1858.

German authors generally consider this amorphous matter to be conjunctive tissue ; concerning the myelocytes,
anatomists are greatly at variance ; but that discussion
would carry us too far and outside of our subject.
To complete a mention of all the elements to be found
in the gray cortex, we should add the very minute fibril
prolongations of the nerve-cells, and also the vessels surrounded by a lymphatic sheath.
A résumé gives as follows :
1. Pyramidal cells.
2. Giant-cells.
3. Myelocytes.
4. Fusiform cells.
5. Amorphous substance.
6. Fibrillary nerve-prolongations.
7. The vessels with their lymphatic sheaths.

SEC. 4. STRUCTURE OF THE CONVOLUTIONS
IN GENERAL.

Let us now examine the relations of these diverse elements and the arrangement of the concentric layers. We
will cite especially from the works of Meynert[1] and Lewis.[2]
We will describe, as Meynert has done, the general type
of the convolutions, following this with the special struc-
tures of this or that one.
A. FIRST LAYER (external).—This layer is formed almost
exclusively of amorphous substance and contains few if

[1] Der Bau der Grosshirnrinde und seine örtlichen Verschiedenheiten, Wien.
Med. Jahrb., 1869; Jour. de l'anat., t. viii., 1872, p. 106 ; t. x., 1874, p. 98.
Stricker's Handbuch für Geweblehre, t. 1., p. 694 et suiv. An excellent résumé
of the latest microscopic researches is to be found in Charcot's Leçons sur les
localisations, 1876, p. 20 et suiv.
[2] On the Comparative Structure of the Cortex Cerebri. Brain, 1878, No. 1,
p. 79.

any nerve-cells. To the unaided eye it appears white, as Baillarger long since observed. Kölliker described and figured it[1] as being especially composed of very fine, intermingling tubes. They are the fibres which Valentin, at the commencement of microscopy, called terminal handles, indeed, according to Kölliker, such would be their appearance. This layer contains few vessels; only the arterioles going from the pia mater to the cortex. This has been observed by Duret;[2] besides, it is a general law in the structure of the nervous system that the parts richest in cells are the ones most abundantly supplied with blood-vessels. The illustration by Luys[3] is very exact.

Meynert and Kölliker regard this layer as conjunctive; that opinion, however, is not general, and Henle, Wagner, Stilling, and Robin think it a true nerve-layer.

According to Meynert, its thickness in different animals would correspond to the cortex entire; one-eighth for man, one-seventh for the monkey, one-sixth for the dog, one-fifth for the cat, one-fourth for the mole, one-third for the calf. According to Lewis,[4] in the sheep its thickness would be about 0.55 mm., the entire thickness of the cortex being about three mill., that is about one-fifth.

Lewis gives to the external layer in the human brain a diameter of from .250 to .340 millimetres, about the fifteenth or sixteenth part of the entire thickness of the cortex, a figure notably different from Meynert's.

According to the same author, there are immediately under the pia mater, cells analogous to the cells of Deiters, adherent to the vessels and so connected with them and the pia mater that in removing the vascular membrane of the encephalon the first layer of the cortex is taken with it.

---

[1] Elém. d'histol., French transl., 1872, p. 399, fig. 206.

[2] Arch. de physiol., t. vi., pl. 6, figs. 2 et 3.

[3] Le cerveau, p. 12, fig. 1 ; the cut is a faithful reproduction of the beautiful photographs which M. Luys has kindly shown me.

[4] Loc. cit., p. 92. See the excellent plates which he has given of the structure of the human convolutions (plate 1), of the cat's (plate 2), of the sheep's (plate 3).

He has seen this both in man and in the sheep ; Major has
noticed the like with the baboon.[1]

Boll and some other authors (Gerlach, Golgi) hold the
outer layer to be mostly composed of Deiters' cells.

B. SECOND LAYER (compact pyramidal).—The second
layer is almost entirely composed of numerous small pyramidal cells crowded against each other. Lewis gives to
these cells a diameter of from 9 to 13 mm. They are a little
larger than the cells of the first layer. There is a little
amorphous matter.[2]

This layer is generally thin. The schematic tables of
Lewis make it the thinnest of all the cortical layers. Its
thickness seems to be tolerably constant, not only in the
various regions of the encephalon, but also with various
animals. Luys considers it the zone of the sensorium
commune ; the opinion, however, is yet a hypothesis.

C. THIRD LAYER (ammonique).—The third layer is also
composed of pyramidal cells, but of a larger size than the
preceding, some of them reaching a diameter of 51 mm.
(Lewis). By some, the layer is divided into two distinct
layers and indeed the most voluminous cells seem to pre-
dominate in the inner portion of the layer. Meynert
thinks the cells fusiform rather than pyramidal, and pro-
poses to name the layer, the Layer of the cornu Ammonis,
as the fusiform cells are the only kind found in the cornua
Ammonis. In addition to these cells, there are in this
layer fasciculi of medullary fibres which rise perpendicularly to the surface of the cortex, forming columns-like
between the groups of pyramidal cells ; the arrangement
is well depicted in the illustrations of Luys and Meynert.

It should be observed that this third layer is thicker
than the first two layers combined. Its appearance differs
greatly according to the region of the brain examined and

---

[1] Observations on the brain of the Chacma Baboon. Journal of Mental
Science, Jan., 1876, p. 498 (see plate 1).

[2] This layer has been well represented by Duval : art. Nerfs, du Dict. de Méd.
et chir. prat., p. 480, fig. 74.

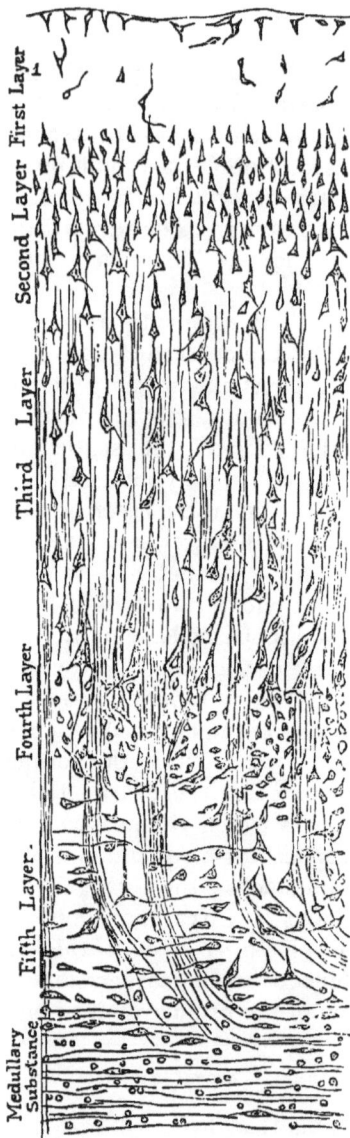

this is chiefly dependent upon the presence or not of the giant-cells of Betz which are only found in the motor centres of the cortex cerebri.[1]

D. FOURTH LAYER (granular).—The Fourth Layer is chiefly composed of myelocytes, lying regularly side by side : the general arrangement of this layer allows it to be compared with the granular layer of the retina, an interesting similarity to which we will have occasion to revert.

E. FIFTH AND SIXTH LAYERS (claustral).—This layer seems to be the most important of all those which combine to form the cortex cerebri. It has a reddish-yellow color (Kölliker) due to pigmentary cells (which are very abundant in aged subjects), and is made up of cells, and

FIG. 4 (after Meynert).—Section of the third frontal convolution of man. (Enlarged 500 times.)
  1. Superficial layer, poor in nerve elements.
  2. Layer of small pyramidal cells.
  3. Layer of great pyramidal cells.
  4. Layer of globular cells.
  5. Layer of fusiform cells.

[1] See Charcot, loc. cit., p. 27.

of fasciculi which form isolated loops, the convexities of which are turned towards the surface of the brain. The fibres composing the fasciculi are at first about 2.6 to 6.7 mm. in size, but at the external portion they become extremely fine, 0.9 to 1.8 mm. in size (Kölliker).

Some authors have thought that they have seen these fasciculi ramify and divide, within the cerebral cortex, between the gray and white substance,[1] but that appears yet doubtful.

It is important to study also the cells. Their form is so characteristic that Lewis terms this inner layer of the cortex the ganglionic layer. They are sometimes stellate, sometimes pyramidal, and sometimes fusiform (large cells of volition of Robin). The fusiform cells are especially abundant at the deeper part of the fifth layer, so much so that it may be viewed as a sixth layer (Meynert). These cells generally present a cylinder-axis (prolongation of Deiters; basal prolongation of Meynert) directed towards the white cerebral substance. Meynert thinks they ought not to be called bipolar, that they probably give off other lateral prolongations more difficult to see.

The dimensions of the fusiform or pyramidal cells are sometimes comparatively enormous, as has been observed by Betz. Lewis gives to the largest, diameters of 126 mm. The dimensions, however, vary greatly, as we will see further on, according to the regions examined. The fifth layer has nearly the thickness of the third. In a rough way, allowing the entire cortex a thickness of three millimetres, the internal or fifth layer would have one millimetre, the third the same, and the first, second, and fourth combined, one millimetre; of course this is but approximative, still it is sufficiently close to the observations which I have made upon various preparations, and to the illustrations of the various authors.

As to a name for this last layer: As Meynert compares

---

[1] Hessling and Schaffner, cited by Kölliker, loc. cit., p. 402.

it to the claustrum (*avant-mur*), and as Vicq d'Azyr, who first closely studied the *avant-mur*, called that the claustrum, I would propose to call the last cortical layer the claustral layer, a term which has the advantage of recalling the discovery of Vicq d'Azyr, and which at the same time does not prejudge its function.

Thus we would have the following layers:

1. External limiting layer.
2. Compact, pyramidal layer.
3. Layer ammonis.
4. Granular layer.
5. Claustral layers, superficial and profound.[1]

The entire gray layers of the brain have a certain thickness, which the medical superintendents of the insane have frequently studied in order to ascertain if there exists a relation between its dimension and the mental condition of their patients. Their results are not yet definite. It is supposed, however, that the greatest depth of the cortex is to be found in the brains of the most intelligent.

H. Major, with the aid of an ingenious instrument (Tephrylometre) has ascertained:

1st. That the thickness of the cortex varies considerably in the different convolutions of the same brain.

2d. That the variations in thickness were not homologous in different brains. Unfortunately these researches have been made upon the insane, and cannot serve in any way to elucidate that very interesting question in normal anatomy.[2]

The designations given are those which we shall henceforth employ; they will avoid fatiguing repetitions and facilitate descriptions.

---

[1] These divisions must not lead to the supposition that the layers are distinctly marked. The limits of the layers are to a considerable degree artificial, and vary somewhat according to the will of the observer, and to the process employed in making the preparation (Luys).

[2] H. Major: A new Method of Determining the Gray Matter of Convolutions. Lunatic Asylum Report, 1872, p. 67.

Sec. 5. STRUCTURE OF INDIVIDUAL CONVO-LUTIONS.

We will first study the motor and the sensorial zones of the cortex, and then the special convolutions (the lobe of the island of Reil, gyrus hippocampi, olfactive bulb, etc).

The anatomical difference in the motor and sensorial zones escaped Meynert, and it was only later, guided by physiology and pathology, that anatomists have been able to discover it.

An important point well brought into relief by Lewis is, that these differences between the zones are never abrupt, it is always gradual, giving place to a transitional zone as it were.

We will first study the type of the paracentral lobule, as it has been studied by Betz.

Morphology, as elsewhere indicated, assimilates that lobule, in a functional point of view, with the two convolutions bordering the fissures of Rolando (asc. frontal and asc. parietal).

The paracentral lobule and the two Rolandic convolutions are characterized by the presence of the giant-cells of Betz. These cells, already described,[1] are found in all the parts which are considered motor-centres.

Charcot[2] remarks that this fact is the more interesting for the reason that these cells are found in all the points considered as motor-centres, whatever may be the mor-phological difference in the convolutions. In the dog they surround the crucial furrow.

Now it has been for a long time known that the motor-cells of the spine are very voluminous, whilst in the sen-sorial regions they are very small, or medium sized. So anatomy as well as physiology indicates a relation be-

---

[1] See page 9, fig. 1.   [2] Loc. cit., p. 29.

tween the size of the cells and their functions. Where
there are motor-centres there are large cells; this is true
of the cortex cerebri as well as of the spinal axis.

To this knowledge Pierret[1] has added the interesting
fact that the dimension of the ·nerve-cell is not only pro-
portionate to its function, but that it also holds relation to
the distance to which it must transmit the motor incita-
tion; that seems proven in case of the spinal cord, and
perhaps an analagous demonstration will be arrived at in
case of the brain.

The large cells are found more especially in the zone
Ammonis (third layer), and in the claustral zone (fifth layer).
Lewis, who has studied the details of their distribution,
gives the following measurement :

	CLAUSTRAL LAYER.		LAYER AMMONIS.	
In man— { Asc. front.	126 m.	55 m.[2]	41 m.	23 m.
{ Asc. pariet.	88 m.	41 m.	51 m.	32 m.
In the cat—Gyrus sig.	106 m.	32 m.	23 m.	13 m.
In the sheep—Gyrus sig.	65 m.	23 m.	18 m.	10 m.

For more details upon this question we refer to the
memoir of the author.

The motor-cells seem to form in groups in very limited
zones, with a determined and constant situation in certain
points of the cortex cerebri; each group appears to be divided into a series of secondary groups, which are the
nests (*nids*) described by Betz.

Except the great cell-layers, the layers in the motor-
zones have no special characteristics; at all events, the
details given by certain authors are too accessory or too
uncertain to be mentioned. I will say the same respecting differences between superior and inferior regions of
the same convolution.[3]

As for the cells of the transition-space between the motor

[1] Comptes rendus de l'Acad. des sciences, 1878, I, p. 1423.
[2] The two figures indicate the large and the small diameters.
[3] See the tables of Lewis.

and sensorial zones, the change is not brusque, and the
cells, though larger than in the occipital lobe, are yet
smaller than in the convolutions anterior to the fissure of
Rolando.   To a certain degree the ascending parietal convolutions may be considered as the transition zone.   In
man, the cells of that convolution measure :

At the top   .   .   .   .   . 88—41 m.
In the centre   .   .   .   . 55—32 m.
At the bottom   .   .   .   . 41—24 m.

To Lewis' opinion may be added that of Betz, who held
that there were two fundamental regions, separated by the
fissure of Rolando.   Anteriorly were the great-cell, and
posteriorly the small-cell convolutions.   Physiology does
not confirm this distinction, for the ascending parietal convolution, back of the fissure of Rolando, manifestly belongs
to the motor-region.   In accordance with Lewis, however,
this convolution can be considered as representing a transitional zone.

OCCIPITAL TYPE.—The sensorial (?) convolutions embrace: 1st, the cuneus ; 2d, the posterior half of the lingual
and fusiform lobules ; 3d, the occipital lobe ; 4th, the first
two sphenoidal convolutions and the marginal fold.[1]

All these convolutions have a structure apparently different from the anterior parts.

Vicq d'Azyr observed a white band (*ruban de Vicq d'Azyr*)
between two gray layers, the whole constituting the gray
cortex of the posterior convolutions.

The correctness of this observation has been confirmed
and perfected by modern researches.

The layer Ammonis (third) is seen to be replaced by two
layers of myelocytes with fewer cells, though relatively
voluminous.   The morphological modification of the layer
Ammonis can be otherwise described in saying that the
original has disappeared and been replaced by the extension of the subjacent granular layer, divided into three
secondary layers, so furnishing eight layers.

---

[1] Charcot, loc. cit., p. 29.

3

Clarke[1] takes the structure of the occipital convolutions as his starting point in the description of the cortex cerebri; he recognizes the two granular zones, but makes no difference between the fusiform and granular cells.

Meynert found, in various parts of the granular layers, cells which he called solitary, and which he deemed very voluminous, though they are not so large as the cells of the motor-regions. Probably they belong to the pyramidal type.

Let us now compare the structure of the posterior sensorial convolutions with that of the retina, a sensitive expansion and in the embryo all but an encephalic convolution.

Owing to the curve of the retinal layer, and to the atrophy of the posterior lamina, which forms the pigmentary layer, the anterior lamina becomes the homologue of the superficial encephalic layer, in such way that the most superficial portion of the convolutions are morphologically represented by the most profound layers of the retina, those which are in contact with the vitreous humor. We can establish a sort of parallel between these various parts, and group them thus; of course, somewhat artificially:

RETINA.	OCCIPITAL CONVOLUTIONS.
Internal limiting.	External limiting.
Layer of nerve-fibres.	Represented by the handle-fibres of Valentin and Kölliker.
Layer of nerve-cells.	Pyramidal layer.
Layer of myelocytes.	External granular layer, which replaces the layer Ammonis.
Intermediate layer.	
Layer of myelocytes.	Internal granular layer.
External limiting layer.— Exists in the retina only as a rudiment.	
Membrana Jacobi.	Claustral layer.

---

[1] Clarke: Proceedings of the Royal Society, London, 1863.

STRUCTURE OF THE CONVOLUTIONS.   23

In brief, from its structure, development and function, the retina can be regarded as an expansion of the sensorial cerebral cortex.

Such, then, are the principal differences existing between the posterior and anterior motor parts of the cortex cerebri, and they may be expressed in two propositions.

A. The motor regions possess giant-cells located in the layer Ammonis and throughout the claustral layer.

B. In the posterior regions the layer Ammonis is replaced by a granulo-glandular layer.

We will now examine the arrangement in some special convolutions. Meynert describes three distinct types:— 1, Found in the fissure of Sylvius and the Island of Reil; 2, in the cornu Ammonis; 3, in the olfactive bulb.

TYPE OF THE FISSURE OF SYLVIUS.—The convolutions surrounding the fissure of Sylvius are remarkable only for the development of fusiform cells, which form a deep claustral layer, better limited here than elsewhere.

Notwithstanding its obscurity, and perhaps because of its obscurity, here arises an important morphological question. The layer of gray substance, extended as a little band between the lenticular ganglion of the corpus striatum and the gray cortex of the island, that which Meynert calls the "*avant-mur*," has been considered as a part of the cortex cerebri, the claustral layer of the three convolutions of the Island. It appears that in the brains of idiots, the white lamina, which separates the insulary convolutions from the *avant-mur*, is absent, and the *avant-mur* really becomes the internal layer of the cortex cerebri (Betz). Now, according to Meynert, the structure of that *avant-mur* is identical with the claustral layer; there are to be found fusiform cells, pressed one against another, a structure widely differing from that of the gray ganglion of the corpus striatum and thalamus opticus. To employ Meynert's expression, the formation belongs, not to the central ganglia of the brain, but to the system of associ-

ation, which always constitutes the internal claustral layer
of the convolutionary cortex.

It is here useless to recall the arrangement of the central ganglia relatively to the *avant-mur*. It is known
that the external capsule, situated externally to the lenticular ganglion, is divided by the gray band (the *avant-mur*)
into two parts, the external one belonging to the cortical
system, the internal, or profound, to the central system.
Besides, the circulation of the *avant-mur* is a part of the
circulation of the convolutions of the island and not that
of the corpus striatum.[1]

I will add that this is not admitted by all authors. Luys
in particular considers the *avant-mur* a dependence of the
corpus striatum. This interesting point demands new researches, but I do not believe the solution belongs to pure
anatomy. Physiology and pathological anatomy must
solve it.

Some English authors have studied the special structure
of the insular convolutions; especially Broadbent[2] and
Major.[3] The latter concludes that there exists no fundamental difference between the layers of the island and
those of the vertex. He has measured the cells of the
island and gives the following figures for the various layers :

    1. Ext. layer, ... 0.008 to 0.012 mm.
    2. Pyramidal layer,   0.012 " 0.02 "
    3. Layer Ammonis, . . . 0.020 " 0.028 "
    4. Granular layer, .   . 0.012 " 0.024 "
    5. Sup. claustral layer, . . 0.020 " 0.024 "
    7. Profound claustral layer, . 0.016 " 0.02 "

The last layer is generally composed of fusiform cells.
The layer Ammonis (third) is the only one of the island
which differs from the same layer in other convolutions.

---

[1] Duret : Anat. Researches upon Encephalic Circulation. Arch. de physiol.,
1874, p. 79.

[2] Structure of the Cerebral Hemisphere (cited by Major).

[3] The Histology of the Island of Reil. West-Riding Lunatic Asylum Med.
Reports, t. vi., 1876, p. 1 et suiv.

The cells contained therein are smaller than in the frontal regions, which perhaps removes the island from being included as a motor-centre.[1]

TYPE HIPPOCAMPI.—The study of the cornu Ammonis and the gyrus hippocampi should be made morphologically as well as microscopically, for it is interesting to show by what modifications Nature has converted a normal convolution into an aberrant one, as in the instance of the cornu Ammonis.

To render the structure of this region comprehensible, I refer to the accompanying schematic plates.

The transverse sections of Luys, best expose the cornu Ammonis. Upon a brain hardened by alcohol, or one quite fresh, let the occipito-sphenoidal lobe be cut in parallel slices commencing at the anterior extremity, and the modifications of the cortex can be observed as it becomes the cornu Ammonis. This is the process long ago recommended by Vicq d'Azyr.[2]

First is seen a little gray band of the outside cortex which becomes invaginated, folds upon itself, whilst the subjacent white substance, at first very large, little by little becomes thinner, as though the lateral ventricle (V) prolonged into the sphenoidal lobe were about to invade it. Gradually this white substance becomes thinner and at last assumes the form of a thin resisting lamina, the wall of the lateral ventricle and the superficial portion of the cornu Ammonis.

This is what anatomists term the alveolus (B). Proceeding posteriorly it will be seen that, upon the internal side of the fold, this cerebral lamina lessens in such way that finally, at the most posterior part, it is reduced to simply a thin layer of white substance, which layer, through constant decrease, finishes by becoming in the lateral ventricle entirely free from the inner side and becomes the

---

[1] A good plate is annexed to Major's memoir.
[2] See Sappey's Anat. desc., t. iii., p. 105 and figs. 457, 458, 459.

PLATE I.

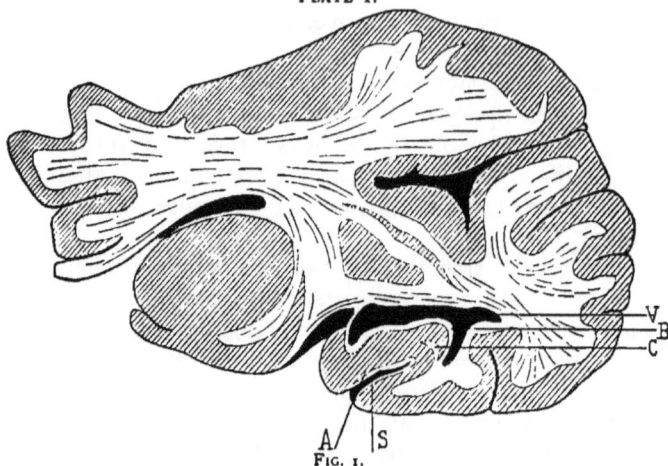

FIG. 1.

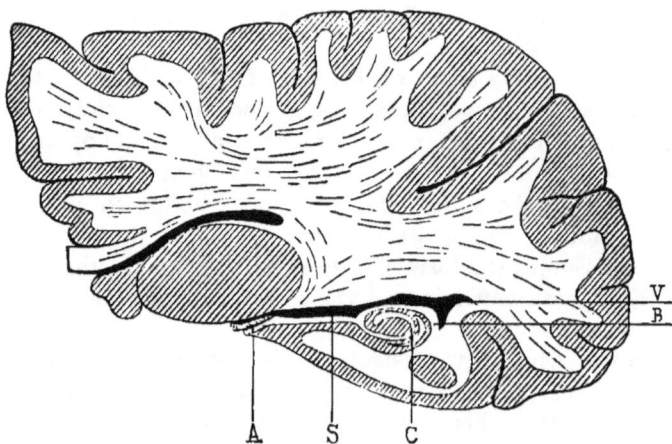

FIG. 2.

EXPLANATION OF PLATES.

(CORNU AMMONIS.)

PLATE I.—Figs. 1 and 2, relations of the hippocampus with the lateral ventricle. The letters are the same as for plate II.

PLATE II.—Figs. 1, 2, 3, 4, 5, 6, 7, 8, 9.   (See next page.)

V. Cavity of the lateral ventricle.

A. Point where the gray substance invaginates to form the hippocampus.

B. Lamina of the deep white substance which tapers down to form the ventricular wall of the cornu Ammonis (cuneus).

S. Subiculum, a white lamina originating in the peripheric gray substance, curving at C so as to form a crosier (bishop's pastoral cross).

To render the figures more distinct, the corpus fimbriatum and choroid plexus are suppressed.

PLATE II. (Cornu Ammonis).

Fig. 1.

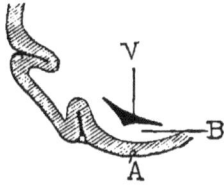

Fig. 2

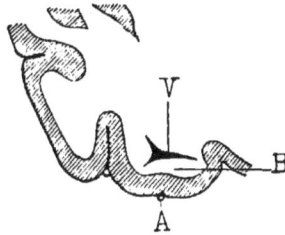

Fig. 3.

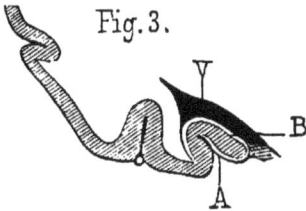

Fig. 4.

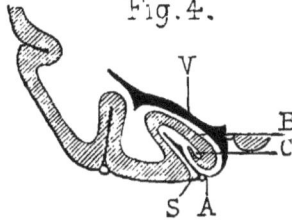

Fig. 5.

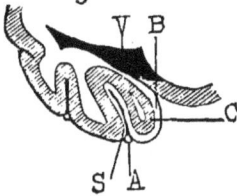

Fig. 6.

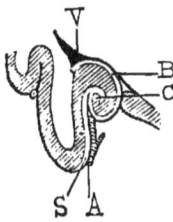

Fig. 7.

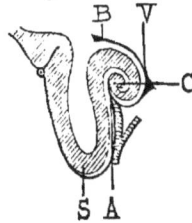

Fig. 8.

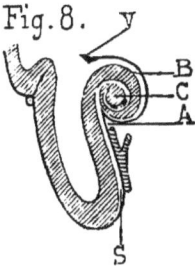

Fig. 9.

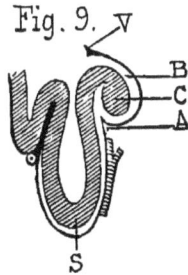

ribbon which anatomists call *corps bordant* (the tænia hippocampi).

Thus the lobule hippocampi is first formed by the fold-
ing of the gray substance and a more or less voluminous
vessel at its base giving off twigs which penetrate between
the two gray laminæ of the convolutionary axis thus
formed. The presence of these twigs in the brains of aged
subjects are of great advantage to the observer.

Following these two layers, it is seen that upon the peripheric surface (now become the central) of one of them
appears a very slight layer of white substance arising from
the gray substance itself (S). This is described by authors
as the *subiculum.*[1]

This white layer is not a dependence of the medullary
white substance of the brain, but a dependence of the gray
cortex cerebri from which it originates.

If this were all, the gyrus hippocampi would be relatively simple ; but a new complication renders the description somewhat difficult. That gray lamina, surrounded by
the alveolus of the internal or ventricular side, and by the
subiculum on the external or peripheric side, coils upon itself like the top of the pastoral staff of a bishop, sometimes
completing a circuit similar to the spiral or helix of a snail.
The gray blade accompanied by these two white laminæ
appears like a twisted, elastic lamina, strongly stretched
and fastened at one of its edges. In the very body of the
gray lamina there is a very attenuated layer of white substance, which can be seen only with a glass and which appears to extend to the extremity of the central lamina.

Finally, to complete this difficult description, we would
add that the thin gray blade seems at its extremity to continue directly with that little reddish band, covered with
projections and called the choroid plexus (*corps goudronné*)

---

[1] It can be very well seen in the schematic figure given by Mathias Duval,
art. Nerfs, du Dict. de méd. et chir. prat., p. 474, fig. 72.—See also in Traité
d'anat. of Sappey, fig. 459, 2—white lamella which separates the fimbriated
body from the gyrus hippocampi.

which, with the cornu Ammonis and the tænia hippocampi, jut into the lateral cavity of the fourth ventricle. Lélut, quoted by Sappey, makes the subiculum pass around the choroid plexus and become confounded with the alveolus.[1]

All these details are somewhat difficult to understand, and I think that they can be followed only by referring at the same time to the plates.[2] (Cornu Ammonis, pp. 26 and 27.)

When the anatomical description is so obscure, it is always difficult to determine the minute structure. We are indebted entirely to Meynert for instructions upon the structure of the gyrus hippocampi, and these instructions are not over-satisfying.

In the cornu Ammonis there exist only large pyramidal cells, and the second layer is lacking.

Also there is an absence of fusiform cells (claustral layer). In other words, the layer Ammonis is largely developed, and unaccompanied by other layers.

Supposing the cornu Ammonis unfolded, a section would discover: 1st, the *subiculum* with very small nerve-cells (homologue of the external layer); 2d, the *lacunal stratum*, with a network of pyramidal prolongations (homologue of the pyramidal layer), but where the pyramidal cells are lacking; 3d, the *stratum radiatum* (homologue of the upper layer Ammonis; 4th, the *pyramids* (homologue of the lower

---

[1] That peculiarity, represented by Sappey (fig. 459, I., 2) has not been figured by Mathias Duval.

[2] Among the atlases which I have been able to consult on this subject (Tiedeman, Foville, Hirschfield, Leuret and Gratiolet, Gall and Spurzheim, etc.) none are exact concerning the cornu Ammonis except the works above cited : Sappey, Anat. descript., t. iii., and Mathias Duval, art. Nerfs, p. 474. I will also add : Wundt, Physiologische Pyschologie, p. 82, fig. 34. The plates of Luys, Recherches sur le système nerveux, 1865, atlas, pl. xxi., figs. 1, 2, 3, 4, 5, 6, are perhaps not very clear. On the other hand, in his photographic atlas the transverse sections (pls. i. to xi.) show the structure of the gyrus hippocampi with a remarkable clearness and exactitude. I regret not having been able to consult the work of Kupffer, eulogistically spoken of by Frey : De structure cornu Ammonis, Dorpat, 1859. Microscopically there is a good cut in Meynert (loc. cit., fig. 236).

layer Ammonis); 5th, the *alveolus* (the homologue of the projection of the ependyma ventriculorum).[1]

OLFACTIVE BULB.—To these varieties of convolutions should be added another, of a special form, so special as to be at first confounded with a nerve; I refer to the olfactive bulb. The olfactive bulb, truly speaking, is as distinct from a convolution as from a nerve; it is an organ by itself, a kind of aberrant type, which can, however, thanks to comparative studies, through morphology and embryology, be classed under the general type of the peripheral gray cortex.[2] Luys rightly compares it to the retina.

The olfactive bulb is, in the embryo, a hollow organ, an olfactive vesicle, the same as there is a retinal and auditory vesicle. With some animals this vesicle remains permanent, presenting an olfactive ventricle, lined with vibratory cells, the bases of which are in rapport with the subjacent nuclei (myelocytes).[3] As for the white root of the olfactive nerve, that should be considered as a commissure of the white substance.

In man, however, this ventricle does not exist. The other constituents of the olfactive bulb have not been rightly studied, and seem to be ill-known, though some labors in this direction have been undertaken.[4]

The external layer seems to be formed of a white substance containing pale nerve-fibres, without myeline, which intermix like a network. Beneath that layer is a zone of gray substance, which contains voluminous multipolar

---

[1] These details will not be understood, except by following the scheme of Meynert, fig. 237, loc. cit.

[2] Broca makes two types of mammifera, according to the volume of the olfactive lobe; the horse and the dog, for example (*osmatiques*), man and monkey, anosmatique or very atrophied. See Revue d'anthropologie, 1878, p. 392 et suiv., Le Lobe olfactive et le sens de i'odorat.

[3] Owjamniskow: Müller's Arch., 1860, p. 54. Walter: Virch. Arch., xxii., p. 241. See also illustration of horse by Broca, loc. cit., Fig. 3.

[4] L. Clarke: Zeitschrift für Wiss. Zoöl., xi., p. 31. Schultze, quoted by Kölliker, Element. d'histol., p. 961. Golgi: Ricerche sulla fina struttura dei bulbi olfactorii, Reggio, 1875. Meynert: loc. cit., t. x., p. 102.

cells and some peculiar bodies, globular in form, first described by Leydig as found in fish, then by Schultze, Golgi, and others under the name of *glomeruli*;[1] they seem to be only a mass of ganglionic cells[2] (Schultze, Kölliker). Meynert's term of *stratum glomerulosum* may be accepted. There is then a second layer, which embraces superficially the glomeruli, and, deeper down, the large nerve-cells, analogous to the cells of Purkinje, in the cerebellum. Between these two ranges of cells are found axis-cylinders, serving probably to connect them. Besides this, these cells give out prolongations, which connect equally with each layer, both of which layers are essentially composed of the fasciculi of the olfactive nerve. In the deep layer there are also pyramidal cells with prolongations, which may be traced to the nerve-cells and to the glomeruli of the middle layer.

Nothing is less elucidated than the structure of the olfactive bulb (olfactive convolution, or, still better, olfactive lobe). New researches are necessary. We will content ourselves with the establishment of two facts—about the only things at present positively known respecting that obscure part of the cerebral structure: 1st, the olfactive nerve-fibrillæ divide and anastomose in a network; 2d, they are in rapport with the nerve-cells and the special cells of the second layer of the olfactive bulb (*Glomeruli* of Golgi).[3]

---

[1] This is what Pouchet and Tourneux call the spheric mass of gray substance (loc. cit., p. 587).

[2] See Meynert, fig. 240. The vessels which go to the glomeruli are shown.

[3] Among the different works undertaken upon the structure of the gray cortex I will cite that of Major, especially interesting in an anatomico-pathological point of view : " On the minute structure of the cortical substance of the brain." West-Riding Lunatic Asylum Reports, 1872, page 41, and various articles published in these reports and in The Lancet (21st July, 1877). The work of Lubimoff ought also to be noted (Arch. de physiol.)

SEC. 6. STRUCTURE OF THE GRAY SUBSTANCE OF THE CONVOLUTIONS OF MAMMIFERA.

Here again the details are few, and information can be found only in the writings of Meynert,[1] Major,[2] and Lewis.[3]

Among various animals there exists considerable difference ; the layers thus far spoken of, however, always exist and are tolerably well outlined, as shown in Lewis' plates of the cat and sheep,[4] the only specimens perhaps which have been given of the convolutions of mammifera. In one of Luys' unpublished photographs of the convolutions of a pig, the external limiting layer—that which is poor in cells, and perhaps of a conjunctive nature—is very thick, and the layers to which the nerve-cells belong occupy not more than two-thirds of the gray cortex ; the claustral layer is seen to be also very thick.

But there is not much interest in stating that this or that layer of cells has greater or less thickness, or different lengths of prolongations in different animals. Besides, it has been little studied, and it is only known that the differences are nearly always confined to the occipital lobes.

On the contrary, the structure of the cerebral cortex of the monkey has a great interest. In comparing it with the human cortex, we may hope in a certain degree to find a reason for the enormous difference of intelligence, or at least without adventurous hypothesis to establish a comparison between the absence or the increase of volume of

---

[1] Stricker's Handbuch, etc., loc. cit.

[2] Observations upon the brain of the Chacma Baboon (Cynocephalus Porcarius, Journal of Mental Science, Jan., 1876, p. 502. This memoir is very remarkable and interesting.

[3] Loc. cit.

[4] Loc. cit., plates 2 and 3.

certain cells, as accompanied by the diminution or increase of intelligence.

Major has studied successively the number and the appearance of the layers, the general character of the nerve-cells and their prolongations, and the number of these prolongations. The following are his principal conclusions:

1st. In monkey and man, the number and the relative dimensions of the cortical layers are nearly identical. There is no difference as to the intimate nature of the cells, and their reactions are identical.

2d. All the layers of cells are similar, save the second one in the frontal convolutions,[1] where in man there abound large cells, whilst in the monkey they are rare. Major supposes, and not without some likelihood, that there is a relation between the volume and number of these cells in man, and the faculty of language special to man, and localized in the frontal lobes. Moreover, it seems that old age and intellectual enfeeblement in man coincides with degeneration of the large cells, so that their paucity of numbers in the Chacma baboon would be a sign of inferiority.

As a rule, the number of the prolongations, and consequently the sphere of action of each cell, is considerably greater in man.

Major has also studied the senile alterations of the nerve-elements of the convolutions of animals, and he insists upon the pigmentary degeneration of the nerve-cells.[2]

---

[1] Major says the second layer, p. 510. It seems, however, that the large cells belong rather to the third layer (Ammonis).

[2] On the Morbid Histology of the Brain in the Lower Animals. West-Riding Lunatic Asylum Reports, 1875.

Sec 7. WHITE SUBSTANCE OF THE CONVO-
LUTIONS.

We now pass to the study of the white substance of the convolutions, and, as with the gray portion, we will not give a general description, but of that only which is immediately subjacent to the gray cortex, and which forms the axis of the convolutions.

It was the ancient idea that the cerebral convolutions were the result of a doubling-up of a peripheric layer, for Sténon strongly protested against that idea and *that error of anatomists.* "There are those who would even have us take the substance of the brain for a membrane."[1]

Long time after that, the same idea was entertained by Gall, who thus expresses himself:—"Each convolution consists of two fibrous layers which are entirely covered by a gray layer of nearly uniform thickness. The laminæ of the cerebellum are formed in the same way. The two fibrous layers, formed by the ascending and diverging fasciculi, are accompanied also by fibres which come from the gray substance, so that each convolution is composed of:—1st, very fine re-entering nerve-fibres; 2d, fibres of the diverging fasciculi; 3d, the exterior envelope of gray substance. It is this arrangement which renders it possible to separate the two layers of fibres without injury to them, and to stretch out a surface and unfold each convolution or duplicature."[2]

That idea of Gall's is perhaps exact, but modern researches certainly are based upon different processes, and notwithstanding that the microscope has furnished better

---

[1] Anatomical Exposition of the Structure of the Human Brain, Winslow, Amsterdam, 1752, t. iv., p. 210.

[2] Anat. et phys. du syst. nerveux, 1810, t. i., p. 299.

grounds for judgment, the question is still environed with uncertainties.

The opinion of Meynert, adopted by Charcot, and the correctness of which seems very probable, is known. It holds the periphery of the brain to be a system of projection. As in descriptive geometry, a projection can be made from the surface of a solid, so the cortex cerebri is the projection of the nerve-fibres contained in the ganglionic centres of the encephalon. To this must be added that these centres furnish a system of condensation, since the number of optico-strio-medullary fibres is much less than the optico-strio-cortical fibres. Besides this, Meynert has described a system of association formed by anastomosing fibres which unite the various convolutions and establish between them a complete and harmonious consensus.

These ingenious views are but hypothetical; and it must be said that, to the present, anatomical observations of the cerebral structure have never passed the domain of hypothesis; we again repeat, that only physiology and pathological anatomy are capable of judging the relations between the convolutions and the nerve-axis.

It should be observed that Baillarger has long since described two rows of vertical fibres in the axis of the convolutions (in the dog and rabbit), crossed by transverse fibres.

The schemes constructed by Luys must also rank as hypothetical. That eminent anatomist has described the relations of the various convolutions with each other and with the central ganglia. In the majority of his conclusions, Luys has certainly arrived at the same results as have pathologists; but as it is the result of pathology which alone carries conviction, we will further along return to what we have to say respecting cortical degenerations.

The knowledge derived by the microscope is scanty and inconclusive.

It is known that the white substance is composed of cylinder-axes surrounded by myeline, and it may be anticip.ted that an attempt has been made to establish a relation between these cylinder-axes and the cells of the cortical layers.

The first point is still scarcely known. We have before seen the opinion of Koschewnikoff; that opinion is very clear and probable, though, as yet, undemonstrated.

According to Gerlach,[1] these nerve-fibres penetrate the nerve-substance, and there, losing their myeline, form a very fine, richly anastomosing, fibrillous network, which fibres have cellular terminations. The existence of this network was not universally admitted. Rindfleisch denied its existence.[2] He maintained that after a series of dichotomous divisions, the nerve-tubes separate into bundles and fibrils; these fibrillæ go, some to the nerve-cells, and others to the myelocytes.[3]

He asks if these nuclei may not be of nerve-origin, thus indirectly putting in doubt the teaching of the German school relative to the interstitial conjunctive tissue.

Prior to these two authors, Kölliker had given the exact relation between the nerve-filaments and the cells.[4]

It must not be supposed that all these observations are easy to repeat, and Vulpian rightly insists, in this question more than in any other, upon a rigid anatomy, which will not be content with approximative results and which will not affirm when there remains a doubt.[5]

According to Kölliker the fibres do not all penetrate directly into the gray substance, there are those which course along the cerebral cortex, forming as it were arches.

In the white substance of the convolutions, there exist

[1] Centralbt. für d. med. Wiss., 1873, No. 18, p. 273.
[2] Centralbt. für d. med. Wiss., 1872, No. 18, p. 277.
[3] Luys had already described the connection of the myelocytes with the nerve-fibres: Atlas, 1865, pl. xx., fig. 5.
[4] See des Elém. d'histol., French trans¹., p. 401.
[5] Leçons sur la phys. du syst. nerv., 1868. For the inter-relations of nerve-fibres, the excellent, though schematic design of Henle may be consulted. Hand. für Nervenlehre, 1871, fig. 203, p. 275.

also nerve-cells which have been described by Mierzejewski[1] and previously by Meynert.[2] These cells are small, from 0.007 mm. to 0.010 mm., and multipolar; their poles are very long; they probably are cylinder-axes. Of the nature of these elements Mierzejewski is yet uncertain. They are confined to the gray substance with which they perhaps ought to be classed.

Their nuclei are very manifest and easily colored by picrocarmate of ammonia.

If we now compare the convolutional structure of the cerebrum with that of the cerebellum, it will be seen that they appear to be constructed upon the same plan. There is only this difference, that the deep layers of the cerebrum are more complex and separated with tolerable distinctness; whilst in the cerebellum, the nuclei, the nerve-cells, and the cylinder-axes are confounded in the formation of the internal or rust-colored layer (Kölliker). Thus we have the following homologue, somewhat schematic:

CEREBRUM.	CEREBELLUM.
Ext. limiting layer. ⎫ Pyramidal layer. Ammonic layer. Granular layer. Claustral layer. ⎭ Gray layer.	⎧ Ex. limiting layer, ⎨ Layer of the cells ⎩ of Purkinge.
	Rust-colored layer. ⎧ Myelocytes ⎨ and nerve- ⎩ cells.
Int. limiting layer.	⎧ Stretched network with quadran ⎩ gular, parallel meshes.

Perhaps the corpora quadrigemina, or at least their superficial portions, should be ranked with the gray cerebral cortex, though it has not yet been done. Although with fishes, birds, etc., the corpora quadrigemina are developed to the extent of veritable lobes, still in the human subject.

---

[1] Loc. cit., p. 202.
[2] Stricker's Handbuch für Geweblehre, 1870, p. 709.

4

they are very reduced; in no case presenting folds and convolutions and their structure is very simple.[1]

Sec. 9. VESSELS OF THE CONVOLUTIONS.

The arrangement of the circulatory apparatus in the surface of the nervous system, apparently so well-known, is really still quite obscure, since upon certain points there is complete discord.

For the historical part, we turn to the work of Duret,[2] where the question is treated of in a very complete manner.

We first observe that the gray substance of the convolutions is much richer in vessels than the white, illustrated by the following cut borrowed from Gerlach. This is a well-known general fact and which applies to the spinal cord as well as to the cortex cerebri.

The gray cortex is so vascular that Ruysch considers it a sort of arterial plexus.

Upon an injected brain may be seen an infinite number of minute branches which fall like a shower of rain, perpendicularly, from the pia mater into the cerebral substance. These are the terminal arteries: they may be divided into two groups: one, very long and voluminous, penetrates through the gray layer and into the white substance; these are the medullary arteries: the other, smaller (cortical arteries) and much more numerous, appear to be distributed in a uniform manner to the gray layer. Between all these branches there is a rich network of anastomoses.

According to Duret, from whom we borrow these details,[3] these arborescences of the gray cortex are responded

---

[1] According to Serras, Anat. comparée du cerveau, Paris, t. ii., p. 277, the surface of the optic lobes are formed of alternate gray and white layers.

[2] Recherches anatomiques sur la circulation de l'encéphale. Historique, p. 343 et suiv., Arch. de physiol., 1874.   [3] Loc. cit., p. 334.

to by similar arborizations from the pia mater, from which it may be inferred that all the vessels ramify before sending branches to the brain.

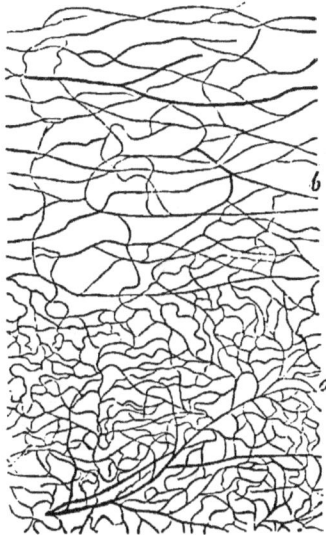

FIG. 5 (After Gerlach).—Portion of injected sheep's brain, showing the difference in the vascularity of the gray (*a*) and white (*b*) substance.

In a section of vascular-injected convolution, the capillary network seems to have four different grades.

Layers. { Pyramidal
Ammonic
Granular
Claustral superf. } Very rich, irregular polygonal network. This is the vascular region of the encephalon.

Deep claustral layer. { Transition network, less rich than the preceding and richer than the subjacent.

White substance. { Network, with meshes three or four times larger.

The veins, like the arteries, may be divided into medul-

lary and cortical. The medullary veins do not follow the
exact course of the medullary arteries.

According to Duret, they communicate with the veins
of the base and of the ventricle of the brain, especially with
the venæ Galeni. The cortical veins are larger and less
numerous than the corresponding arteries.[1]

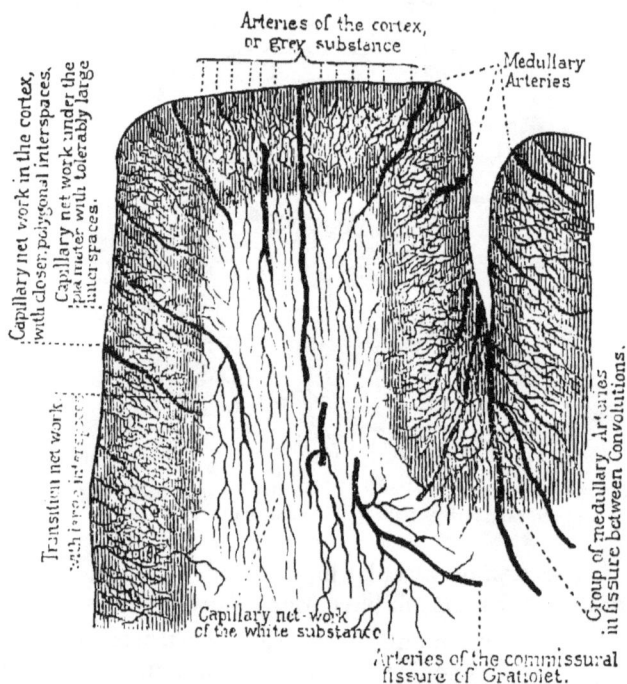

FIG. 6—Circulation in the convolutions (after Duret).

It seems probable that there are no canals of communication between the arteries and veins of the gray cortex
other than the capillaries.

---

[1] See Duret (loc. cit., p. 338), who has studied the question with minute care.
Charcot (Leçons sur les Localisations, 1876, p. 54) has shown that Hubner's
researches on the subject were absolutely simultaneous and not posterior, having been communicated upon the same day (7th Dec., 1872), a portion to the
Biological Society, and a portion to the Centralblatt für med. Wissensch.

Regarding the pia mater the question is still under discussion ; the objections of Duret to the theory of Ecker
and Sucquet, which admits derivative canals, seems to me,
however, very serious.   As with all questions in dispute,
the subject requires new researches.[1]

An interesting thing, from various points of view, is the
relative independence of the different vascular territories
of gray cortex cerebri.[2]

Determined more especially by pathology, Duret and
Charcot hold that there are two very distinct regions in the
distribution of the cerebral arteries : the cortical region and
the central (*corps opto-striés*).   " These two systems," says
Charcot, "although common in origin, are entirely inde-
pendent of each other, and at their peripheries they have
no point of communication."

Not only is there no communication between the cortical and medullary systems, but the various divisions of the
cortical system do not intercommunicate except by very
fine capillaries, so that the independence of the several
cortical regions is nearly complete.

Duret says [3] " that in man, the dog, and the rabbit, there
are three separate regions, each furnished by a special
artery ; the anterior cerebral arteries supply the frontal
lobes, the sylvian (median) arteries the convolutions about
the fissure of Rolando, and the posterior cerebral arteries
the occipital lobes."

Cadiat opposes various objections to the views of Duret,[4]
the most important one seems to be that in making even a
gentle injection into any branch of the arterial hexagon
(circle of Willis), the entire lobe becomes injected, but it
must be observed that Duret and Cadiat do not absolutely

---

[1] According to Heubner and Cadiat, there should be anastomoses between the
veins and arteries of the pia mater.

[2] See fifth and sixth Lessons upon Les Localisations du cerveau, by Charcot.
Part I., 1876, p. 53 et seq.

[3] Revue des Soc. méd., t. x., 1877, p. 428.

[4] Bull. de la Soc. de Biol., 1876, p. 342.

differ, for Duret does not deny anastomoses of the arterioles of either the pia mater or the brain; he only contends
that their diameters do not exceed one-fourth of a millimetre, whilst Cadiat and Huebner allow them one full millimetre. It then is but a difference of degree, and the question is evidently difficult to settle, for an arterial injection
always distends the capillaries.

In short, to Duret the anastomoses are not important,
to Cadiat they are. The matter is still undecided; the
mode of operation and the kind of injection employed have
greatly to do with the affair. I would add, however, that
the exact limitation of cerebral infarctus and embolism
lends a certain degree of support to Duret's theory.

I will say a few words only concerning the structure of
the cerebral arterioles. It is known that Robin has proven that the cerebral capillaries are surrounded with a
lymphatic sheath.[1]

That sheath is studded with ovoid nuclei; it has a thickness of 0.001 mm. to 0.002 mm., and completely surrounds
the vessel in such a way that two concentric tubes may
be seen, each containing different liquids.

The internal tube contains red globules; the external,
leucocytes. As yet it is by no means proven that these
canals are lymphatic, inasmuch as they have never been
injected or traced immediately to a ganglion.

German anatomists have met this proposition with a
multitude of conjectures (cellular-perilymphatic spaces,
cellular perivascular sheaths, etc.). Some histologists have
even thought these canals to be pathological alterations
(Kesteven[2]). But as we have not general anatomy under
consideration, it suffices to note the existence of a lymphatic sheath to the blood-capillaries, as if the cerebral tissue
were too delicate to endure the immediate contact of the
blood.

---

[1] For the complete bibliography see Riedel : Die Perivasculäre Lymph-
räume, Arch. für mik. Anat., t. xi., 1875.
[2] Brit. Med. Jour., June, 1874, p. 840.

Sec. 8. DEVELOPMENT OF CONVOLUTIONS.

Notwithstanding the considerable number of authors who have written upon the anatomy of the nervous system, the development of the convolutions, structurally, not morphologically, has received the attention of but few.

Until the third month, as Tiedemann [1] has observed, the surface of the hemispheres is smooth, without fissures or convolutions. According to both Tiedemann and Duret,[2] the first cerebral folds appear at about the third or fourth month. The first fissure is not that of Rolando or Sylvius, as the plates of Leuret and Gratiolet [3] would seem to to indicate, but the external perpendicular fissure.[4]

Until the fifth or sixth month, there is scarcely any development of true convolutions. Gratiolet has represented the brain of a five-months and a half fœtus,[5] which is entirely analogous to an accompanying illustration of the brain of a stupid monkey. There are incisions, shallow fissures, outlining of lobules, or groups of lobules; but there are no true convolutions, intermixed, compli. cated, and with blunt angles as in adult brains. Even with new-born children [6] the convolutions are very simple, so that they continue to form after birth; the sulci deepen, new marginal gyri appear. In a word, the convolutions seem obstructed, *mal-arrêtés*, to use Parrot's expression.[7] It is probable that these exterior differences correspond to

[1] Anat. du cerveau, trad. de Jourdan, Paris, 1823, p. 36, pl. 1.
[2] Bull. de la Soc. de Biol., in Gaz. méd., 1877, p. 172.
[3] Anat. comp. du syst. nerv., Paris, 1839-57, pl. xxix., figs. 3, 4, 5.
[4] See also Duval, in Gaz. Méd., 1877, p. 161.
[5] Mém. sur les plis cereb. de l'homme et des primates, p. 82. Atlas, pl. xi., figs. 1, 2, 3.
[6] Leuret and Gratiolet, loc. cit., pl. xx., fig. 2.
[7] Etude sur le ramollissement, etc., Arch. de Phys., 1873, p. 63.

differences of structure, but this point does not seem to have been studied.

The first attempt seems to have been made by Tiedemann,[1] who wrongly supposed that the cortical substance was secreted by the pia mater upon the surface of the brain after birth. But Baillarger[2] has shown that, in the human fœtus of four or five months, there already exists at the periphery a very thin layer of gray substance, showing several concentric zones, alternately opaque and transparent.

According to Duret,[3] from the seventh to the ninth months, whilst the convolutions are forming, there appear, on the one hand, nerve-tubes which reach the superficial portion of the cerebral hemispheres, and, on the other hand, the nerve-cells in the cortex. It is not until this period also that the arteries develop.

Duret considers the coincident development of the nerve-cells, vascularization, the ascent and development of the peduncular expansion and the appearance of the convolutions to prove that the cortex does not acquire functional properties until towards the end of fœtal life. Physiology has arrived at a similar conclusion.

Lubimoff (of Moscow) has arrived at analogous results. The nerve-cells of the cerebrum and cerebellum develop last, whilst the cells of the spinal cord, and especially those of the great sympathetic, develop much more rapidly.

Mierzejewski[4] holds that, in the white substance of the newly-born, among the axis-cylinder, may be found amœboid cells (which are colored black by osmic acid), polygonal cells and small, ovoid, flat cells. To explain the rôle which he assigns to these elements, it would be necessary to enter into the history of the development of nerve-tissue.

---

[1] Loc. cit., p. 87.

[2] Mémoire sur la formation des centres nerveux. Jour. l'Esculape, 1840, et Mém. de l'Acad. de Méd., 1840, p. 150 et suiv.

[3] Bull. de la Soc. de Biol., in Gaz. Méd., 1877, p. 172.

[4] Loc. cit., p. 202.

Respecting the development of the gyrus hippocampi, Duret, who made observations upon a four-months' fœtus, says that it is simply a folded convolution, which projects underneath the corpus callosum. Anteriorly this convolution disappears, little by little, but posteriorly it remains, as before observed, and folds back upon itself. In the large work of Mihalkowitz [1] are to be found details, too long and unimportant to engage our attention, and the plates are also difficult to comprehend. Flechsig, in his great work on the nervous system, does not dwell upon the stratification of the layers : he describes the relations of the various fasciculi of nerve-fibres, and thinks that the white substance is developed very late in the embryonic convolutions; in other words, that Meynert's "*système d'association*" becomes established only at birth.[2] This is pretty much Duret's conclusion.

There are two other interesting facts to be added to the little which is known upon the development of convolutions. First, the existence of "transitory convolutions" found in the embryon of four or five months, and which disappear towards the seventh or eighth months.[3]

Second, Major,[4] who examined the cortical nerve-cells of an eight-months' fœtus, reports the almost complete absence of cellular prolongations, so that the cells have the appearance of round cells. Perhaps the nerve-cells (motor or sensorial) only develop with the commencement of functional action. It also appears that the nerve-prolongations become lost in advanced age.

It would be an interesting study to ascertain at just what time the large motor-cells first appear. I have not had

---

[1] Entwickelungs-Geschichte des Gehirns, Leip., 1877, p. 145 et suiv., pl. xx., xxi.

[2] Die Leitungsbahne im Gehirn und Rückenmarke des Menschen. Leip., 1876, in Jahresber. für Anat., 1876, p. 275.

[3] Milhalkowitz, loc. cit., p. 144. Ecker : Arch. für Anthrop., 1868, t. iii. His : Entwickelungs-Geschichte der Grossgehirnhemisphären. Sitzber. der Nat. Gesellschaft in Leipzig, 1874, p. 1.

[4] Loc. cit., Jour. of Mental Science, p. 511.

time to undertake it. I would observe, however, that, in a cat one month old, I have, in company with M. Tourneux, noticed that there was, even at that age, a difference between the anterior and posterior cortical cells, the latter being a little smaller. Betz says that giant-cells do not exist in the new-born.

SECOND PART.

PHYSIOLOGY OF THE CONVOLUTIONS.

ANATOMICAL INTRODUCTION.

Although the topography of the cerebral convolutions is known, still a short anatomical review may not be useless and will avoid repetition and confusion.

The human brain is composed of four distinct portions.[1] The anterior plane includes the frontal lobe; the middle plane includes, superiorly the parietal lobe, inferiorly the sphenoidal, the posterior plane the occipital lobe.[2]

To these we will add other lobes of secondary importance, at least to man: the olfactive, very reduced; the gyrus hippocampi, impossible to well determine; the lobe of the Island of Reil, deeply hidden in the depths of the fissure of Sylvius; the lobe of the corpus callosum, and the gyrus angularis.

In the lobes are to be distinguished lobules and convolutions. The lobules are but topographic regions, whereas the convolutions have a real anatomical existence.

Between the various lobes are fissures.

Between the frontal and parietal lobes is the fissure of Rolando; between the parietal lobe and the gyrus angularis is the occipital fissure (or perpendicular) ; between the fronto-parietal lobe and the temporal, is the fissure of Sylvius.

Between the fronto-parietal lobe and the lobe of the corpus callosum lies the calloso-marginal fissure.

Between the occipital lobe and the gyrus angularis lies the fissure calcarina.

---

[1] The anatomy and morphology of the convolutions belong entirely to French science ; Vicq d'Azyr, Rolando, Foville, Leuret, Gratiolet, Broca.

[2] We adopt the terms employed by Broca. Mém. sur la nomenclature cérébrale (Revue d'Anthrop., 1878, p. 193).

Between the convolutions are furrows (sulci).

The frontal lobe is composed of three convolutions: first or superior, second or middle, and third or inferior.[1]

The inferior portion of the frontal lobe is known as the orbital lobule; upon which may be observed three orbital volutions, prolongations of the frontal volutions.

Surrounding the fissure of Rolando are two important convolutions; anteriorly the ascending frontal convolution

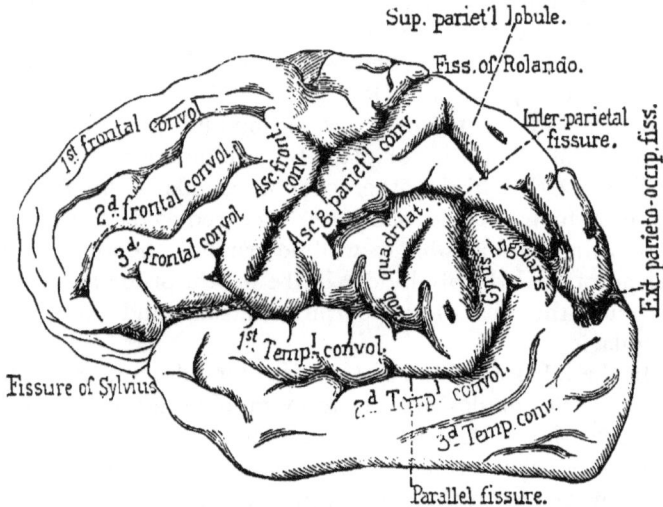

FIG. 7.—Convex surface of a hemisphere of the human brain (parietal lobe partly schematic). (Foville.)

or better, pre-rolandic, which seems to give three prolongations, that is, the three frontal convolutions. Posteriorly is the ascending parietal, or better, post-rolandic.

The union of these two convolutions on the internal face of the hemisphere forms the paracentral lobule.

The temporal lobe includes the first, second, and third temporal convolutions, Fig. 7. The fissure separating the

---

[1] As remarked by Charcot, it would be proper to make an exception and call the third frontal convolution the convolution of Broca.

first and second temporal convolutions is the parallel fissure, which terminates in the parietal region, called gyrus angularis.

The parietal lobe is divided into two parts by the interparietal fissure. This gives a superior and an inferior parietal lobule.

The occipital lobe embraces the first, second, and third occipital convolutions. Inferiorly, the temporal and occipital lobes seem confounded, so as to furnish a first and second temporo-occipital convolution.

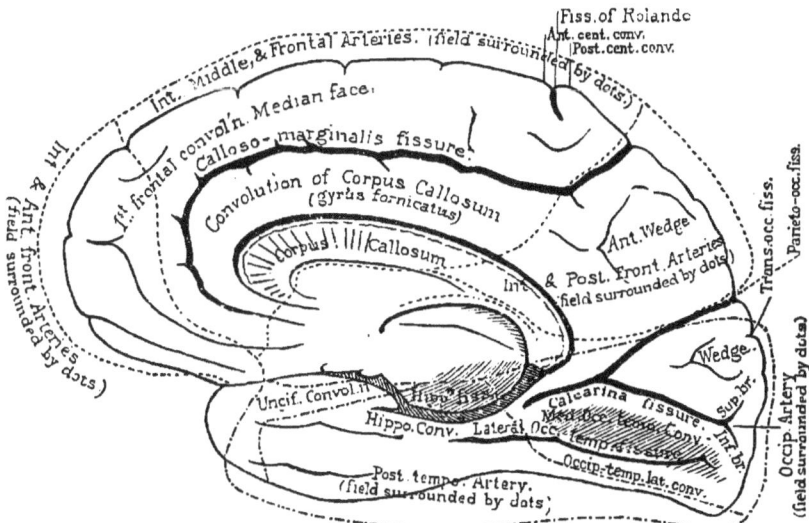

FIG. 8.—Internal face, right hemisphere of human brain (Ecker.)

This nomenclature applies equally to man and monkey : but as one cannot experiment upon man,[1] and as physiologists rarely have monkeys at command, it is of special im-

---

[1] There are exceptions to all rules. Bartholow (Revue des Scien. Méd., t. iv., 1874, p. 65), gives some interesting experiences with one of his patients. He plunged needles into different parts of the brain, passed electric currents through them, and watched the results. The patient died two days after, but the needles had nothing to do with the death!

portance to have a full knowledge of the dog's brain and its comparative parallelism with man's. Thus pathological facts in man and physiological experiments upon the dog can be systematically compared, so that a very good idea can be obtained of the functions of the various convolutions according to their location.

The configuration of the dog's brain, as first shown by Gratiolet, has no direct relation with man's, at least the homologies are not at first seen, and the type is different.

The fissure of Sylvius exists, but not that of Rolando; or at least it is replaced by a much more anterior fissure, called the crucial (B, Fig. 9).

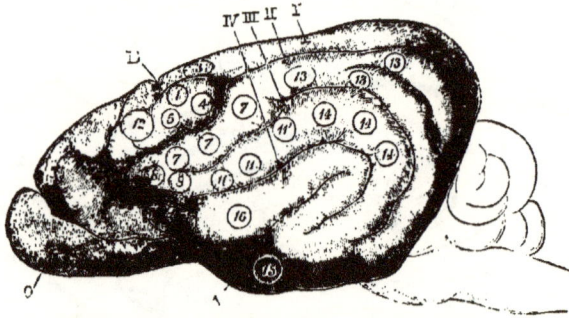

Fig. 9.— Right hemisphere of dog's brain (after Ferrier).

A, Fissure of Sylvius ; B, crucial fissure ; O, olfactive bulb; I., II., III., and IV. represent respectively the first, second, third, and fourth convolutions.

The crucial fissure crosses the inter-hemispheric fissure at right angles, giving the appearance of a cross. It is the same with the cat, only that the crucial fissure is still more anterior than in the dog.

About the crucial fissure is a convolution which seems to respond to the pre- and post-rolandic convolutions in men.

The olfactive lobe is greatly developed, the frontal scarcely at all.

HISTORICAL INTRODUCTION.

The brain of man differing greatly from that of animals, it can be foreseen that experiments upon animals will not give results exactly applicable to man. This fact must be specially emphasized, as it is a very important and commanding one.

Compare the human blood with that of the sheep or fish. As these different kinds of blood have the same functions (absorption of oxygen and nutrition of tissues), have the same general chemical constitution and a very analogous anatomy, it will suffice to examine the function of the sheep's or fish's blood in order to understand the function of human blood.

The same for other tissues and organs; the kidneys for example, or the muscular tissue. The same also for certain nervous functions, innervation of the heart or blood-vessels: that observed in animals can serve in human physiology.

But for the encephalon, and especially for the cerebral convolutions, this identity no longer holds.

For example, if Cuvier's brain be compared to that of a dog, it will be seen that the anatomical constitution and the physiological function are very different; and therefore the conclusions of physiological experiments upon the brain of a dog cannot be applied exactly and absolutely to the human brain.

A very important reserve, however, should always be made; that is, that these differences are *quantitative*, not *qualitative*. To explain: The reaction of the cortex cerebri to excitants should be, and really is, identical in man and dog; it would be more or less marked, more or less extensive in one case or the other; but the functional essence would remain identical, as is the case with the blood, the spinal cord, the nerves, and heart.

If we observe that the cortex cerebri is both motor and sensorial in the dog, that would be sufficient to allow us to affirm the same in case of man.

The question can be stated still more clearly. In the sigmoid gyrus of the dog, we find a motor-centre for the fore-legs; that permits us to say that it also exists in man. But can we venture to say that this motor-centre in man is near the fissure of Rolando, and that the ascending frontal convolution corresponds to the sigmoid gyrus of the dog? Should we dare to say that there exists for the anterior member one motor-centre and not two, or three or four? That would exceed the limits of legitimate deduction. From the fact that there are motor-centres in the dog, we may conclude that there are also motor-centres in man. From the fact that intelligence in the dog depends upon the convolutions, we may conclude that it is the same with man; but we can go no further.

Applications to human physiology would very soon be limited were there not another precious source of knowledge; that is, pathological anatomy and physiology.

Pathology and physiology do not antagonize; they are two branches of the same science, biology, and they should afford mutual light. Though physicians are too often ungrateful to physiologists, the latter should not return ingratitude, and disdain the countless contributions which are scattered through medical lore upon the subject of the functions of the cerebral convolutions. Good observations equal good experiments, and we are resolved to profit largely from the valuable gifts which pathological anatomy offers to the study of the cortex cerebri.

However, in the study of cerebral convolutions, the two sciences differ in their points of advantage and disadvantage. Physiology has these two advantages:

1st. The experiment can be repeated as often as desired.

2d. The conditions of the experiment can be determined, a thing necessary to the value of the phenomenon.

The advantages of pathology are also considerable.

1st. The lesions are upon the human subject, in whom the encephalon differs much from that of animals.

2d. The lesions are always better limited than in physiological experiments.

3d. The symptoms are studied for a much longer time, (and probably with more care).

4th. The subject can describe his sensations.

Thus we think that physiologists should profit from the results of pathology, and that in all physiological study we should highly estimate medical observations made upon man.

The ancient authors, especially Galen,[1] had very incomplete ideas. Galen called the gray cortex of the brain *cpcncranis* (ἐπέγκρανις), after Erasistratus, and the folds of the brain (ἕλικες). Erasistratus thought the human epicranium more complex than animals, for the reason that he has more intelligence. To this, Galen offered a rather worthless argument: "Asses," he said, "have a very complicated encephalon, whereas their *imbecile* character would exact an encephalon quite simple and free from variations." Whatever the ground of that strange idea respecting the imbecility of the ass and the complexity of his convolutions, it may be seen that, even in ancient times, the brain and even the cortex cerebri was considered as the seat of intelligence. To the quotation already given Galen added this curious sentence, showing the admirable sagacity and prudence of this genius:

"To refrain from speaking of the substance of the soul, when speaking of the structure of the body which contains it, is impossible ; but if this is impossible, it is possible to turn promptly away from a subject upon which we should not dwell." This is the programme which we shall attempt to follow.

Galen also noticed that the brain is insensible.[2] This

---

[1] De usu partium, viii., 13 Ed. de Daremberg, t. i., p. 563.
[2] Cité par Longet, loc. cit., p. 640.

important fact has since been observed by other writers, and seems now well attested.

From Galen to the commencement of the present century, only scattered facts can be produced.

Surgeons of the seventeenth and eighteenth centuries thought that lesions of the cortex cerebri might produce paralyses, and to remedy this, they had recourse to trephining; but their opinions are very confused. Concussion, and above all compression, played the principal part in their theories. It should be remarked also that, in the cases observed, it is rare that the lesions are exactly located. Concussion, consecutive hemorrhage, and encephalitis quickly extended to all parts of a hemisphere.

Lorry,[1] however, gave some very exact experiences and stated that the cerebral pulp was insensible.

Haller,[2] with some important restrictions, expresses very nearly the same opinion. He says that it is necessary to go deeper than the cortex cerebri in order to provoke movements or sensations, and that the medulla of the brain is the sensitive portion.

"*Non ergo videtur aut sensum in cortice cerebri exerceri, aut plenam perfectamque causam motus musculosi in eo habitare, cum præterea plurima experimenta demonstrent, profundo demum loco, et a cortice cerebri valde remoto medullam lædi oportere, ut convulsio superveniat.*"[3]

I dwell upon the ancient ideas only because they have not until the present ever been disturbed, the doctrine of Galen, Lorry, and Haller having held sway until 1870.

The importance of cerebral convolutions as related to the intellectual faculties, though suspected by physiologists and medical practitioners,[4] was especially brought to light by Gall. Gall's merit was not the invention of an absurd

[1] Mém. de l'Acad. des Sciences (Recueil des savants étrangers, 1700, t. iii., p. 352).
[2] Elementa physiologiæ, t. x., p. 312 et suiv., lib. x., § xx. Num cerebri medulla sentiat.
[3] Same, p. 392, § xxiii.
[4] Van Swieten, t. iii., p. 264; t. ii., p. 604, Boerhave, etc.

theory, but the proving, by comparative anatomy, and by the study of the brains of idiots and the insane, that intelligence is a function of the convolutions.

He was followed in this direction by Flourens, one of the most vigorous opponents of phrenology. This celebrated physiologist[1] demonstrated beyond dispute that the encephalic nervous system is the seat of intelligence, the origin of sensation and motion. From this time, that which was previously but a supposition or a feeling, became a positive acquisition to science, based upon firm, unshakable proofs.

We must turn to the experiences of contemporaneous physiologists for new facts respecting the physiology of the convolutions; but from Flourens to the present time (evidently guided by his labors), other authors, zoölogists, and medical practitioners have furnished a large number of interesting facts well calculated to illuminate physiology.

It is somewhat remarkable that the three most important observations occurred in the same period, and are due to three French savants.

1st. Desmoulins[2] demonstrated by comparative anatomy that the number and perfection of the intellectual faculties are in direct ratio to the number and depth of the cerebral convolutions.

2d. Calmeil,[3] in an admirable series of observations. proves that with the insane, especially with general paralytics, the alterations are in the gray portions of the convolutions, and that therefore there is a direct connection between mental disturbances and lesions of the cortex cerebri.

3. Bouillaud, the illustrious dean of French medicine,[4] demonstrated by a great number of pathological facts that language is located in the anterior lobes of the brain.

[1] Mémoires lus a l'Institut, 1822, 1823. Recherches expérimentales sur les propriétés et les fonctions du système nerveux, 1st Ed. 1824, 2d Ed. 1842.
[2] Anatomie du système nerveux des vertébrés, 2d partie, p. 606, Paris, 1825.
[3] De la paralysie chez les aliénés, Paris, 1826.
[4] Traité de l'encéphalite, Paris, 1825, p. 279.

From 1825 to 1861, the physiological history of the cortical system remained absolutely stationary. In 1861, Paul Broca, in a remarkable mémoire[1] made it apparent that language was not only located in the anterior lobe, but that it was confined to a special convolution, and to the posterior portion of that convolution (third convolution of left frontal lobe).[2]

About this time, the conformation of the human brain began to be looked upon as an orderly development instead of a result of chance. The researches of Foville, and those of Leuret, Gratiolet, and Broca (1855–1865), established the constancy of the form of the convolutions.

Excepting some isolated instances, the profession did not admit the localization of cerebral functions, beyond that concerning language.[3]

The celebrated researches of Fritsch and Hitzig[4] terminated these hesitations and established the motor-power of the cortex cerebri.

Whatever interest attaches to subsequent labors, it must be recognized that the first work of Fritsch and Hitzig contains all that is essential upon the question.[5] They have shown:

1st. That there are motor-centres in the brain (excitable by electricity), and again that certain portions are not thus excitable (p. 311).

2d. That the points where excitation affects certain

[1] Sur la siége de la faculté du langage articulé, avec deux observations d'aphémie. Bull. de la soc. anat., 2d série, t. iv., 1861.

[2] It is but just to say that the relation between language and a lesion of the left hemisphere had been noted by Dax in 1836 (See Dax fils, Gaz. hebd., 28th April, 1865.

[3] Hughlings Jackson, 1868.—Prevost, a pupil of Vulpian, says in his inaugural thesis, p. 140, "crossing of the eyes may be observed in cases of superficial lesions of a hemisphere." 1868.

[4] Ueber die Electrische Erregbarkeit des Grosshirns. Arch. für Anat., 1870, 28th April, p. 300–332.

[5] It may be interesting to read the polemic acerbity of Hitzig respecting Carville and Duret. Gaz. méd., 1875, Feb. 6th, and Arch. für Anat., etc., 1875, p. 428 et suiv. But Hitzig defends a cause not attacked.

groups of muscles are very precisely limited to a small area of the cerebral surface (p. 311).

3d. That the results are more regular with the direct than with the induced current (p. 316).

4th. That the removal of a localized cerebral region with the scalpel will produce paralyses (p. 328).[1] The question of motor functions of the convolutions has been considered by numerous writers. Among all the newly demonstrated facts there are four of some importance:

A. There are no motor-centres in the newborn (*Soltmann*).

B. Excitation of the white layer below, gives the same results as exciting the gray substance above (*Dupuy, Carville, and Duret*).

C. The motor-centres of the limbs are also vaso-motor, excito-secretor (*Bochefontaine and Lépine*), and sensorial centres (*Vulpian*).

D. There are sensorial centres in the occipital convolutions (*Ferrier*).

In a medical point of view, the researches of Charcot and his pupils have established upon a firm basis the theory of localization, thus medicine has lent to physiology a proof which was perhaps necessary before admitting the existence of motor-centres.

We will now study the physiology of the convolutions under two aspects; the properties of the gray cortex, and its functions. Indeed, in the physiology of organs there

---

[1] Some remarks, more or less precise, do not establish the title of priority to a discovery. A single observation of Griesinger (see Bernhardt, Arch. für Psychiatrie, iv., p. 480), or a remark of Eckert (Exper. Phys. des Nervensystems Giessen, 1867, p. 157) are not sufficient. In this way Broca should be cited who says, in 1861 (Bull. de la Soc. d'anthropol., p. 318): "The posterior convolutions differ notably from the middle and anterior convolutions. The principle of cerebral localizations is established both by physiology and pathology, the latter showing the independence of the functions, and also by anatomy, which shows the diversity of the organs." Neither should the persevering labors of Hughlings Jackson be forgotten. But these do not take from Fritsch and Hitzig the incontestable right of priority.

should always be distinguished a state of repose and one of
activity ; they may be termed, respectively, static and
dynamic.

According to this division, we will first examine the excitability of the convolutions, their electric condition, and
their nutrition.   In the second chapter, we will consider
their relations and functions in the organism as regards
motion, sensation, and intelligence.

FIRST CHAPTER.

PHYSIOLOGICAL PROPERTIES OF THE CON-VOLUTIONS.

A.—EXCITABILITY.

The apparent inexcitability of the cerebral surface, the inability of chemical, mechanical, or other agents to pro-voke motion is something which has attracted the attention of all observers from Galen to Lorry. The experiment of Fritsch and Hitzig, however, renewed attention to the sub-ject, and to-day the question stands, Is the gray substance inexcitable?

Writers generally hold that mechanical and chemical agents are incapable of exciting motion.[1]

Respecting the chemical or mechanical inexcitability of the gray substance, however, authors are not in accord. Brown-Séquard[2] has recently made some interesting experi-ments in this field. According to him, mechanical and especially thermic excitations of the cerebral surface pro-duce, at least temporarily, the same effects as a section of the cervical sympathetic nerve of the side corresponding to the excitation. These phenomena would be as complete as after section of the sympathetic; moreover, that action would not be produced except after excitation of the right hemisphere. Eulenberg and Landois have noticed analo-gous phenomena. In applying sea-salt to the cerebral sur-face, they found, first, a lowering of temperature (excitation) followed by an elevation of temperature in the fore-limbs,

---

[1] Nothnagel, cited by Dupuy (Lond. Times and Gazette, No. 1410, 1877), found the rabbit's brain mechanically excitable with a needle. Dupuy wittily added that Nothnagel's rabbits differed from those which could be obtained in France and England (see Nothnagel, Virchow's Arch., lviii., p. 420).

[2] Arch. de physiol., 1875, p. 854.

which they attributed to the destruction of the gray cortex.[1]

We cannot tell how far these effects are attributable to the action of the cerebral cortex. It is likewise doubtful if movements of the limbs, such as those following galvanic excitation, can be produced by mechanical or chemical excitations to the brain, or if either the gray or white cerebral substance can be excited by those agents.

The question of excitation by galvanism is still more difficult and more undecided.

When certain regions of the gray cortex are excited, the sigmoid gyrus in the dog for example, either by a moderate direct, or an induced electrical current, movements of the limbs follow, and it may be concluded that the gray substance has been excited and has produced them.

The conclusion, however, would be rather superficial, and we are indebted to Dupuy[2] and Carville and Duret[3] for having shown that the electric current diffuses at the base of the brain, and produces excitement of the white substance.

In placing at the base of the brain the nerve of a galvanoscopic frog, Dupuy has seen electrization of the sigmoid gyrus produced a movement of the paws as before mentioned.

Carville and Duret have also studied, with the galvanometer, electric diffusion upon the brains of both dead and living animals, and they have noticed very feeble induced currents to extend themselves from one to another point of the periphery, extending at the same time also a certain distance down into the white substance.

It always seemed to me that experiments with the galvanometer were of no special importance, and proved very

---

[1] Berl. Klin. Woch., 1876, Nos. 42 and 43, also Virchow's Arch., t. lxviii., p. 245.

[2] Thèse inaugurale, Paris, 1873. Examen de quelques points de la physiologie du cerveau, pp. 23, 26.

[3] Bull. de la Soc. de Biol., 20th Dec., 1874, p. 374.

little, for the reason that the instrument is so sensitive that it shows currents of diffusion almost everywhere. In electrizing the right arm, there will be produced an electric state of the left arm, and a current which will produce an enormous deviation of the galvanometer.[1]

It is preferable to use the sciatic of the frog, which has a very sensitive reaction, and evidently quite sufficient, for when the nerve will not react, the cerebral substance cannot, as it is less easily excited than the galvanoscopic leg.

Very moderate electric currents induce movements in the legs of a dog; but the galvanoscopic leg will not be excited, provided it is placed at a sufficient distance from the cerebral points excited, say one or two centimetres or more; consequently, the electric current, though diffused physically, is not physiologically so diffused but that the excitation may be limited to certain well-defined portions of the brain.

Moreover, a very simple experiment demonstrates that the current may be localized in certain points of the periphery, since either negative or positive effects can be obtained at will by exciting two points separated by intervals of not more than one or two millimetres [2] (Rouget).

I have made analogous experiments with the galvanic limb and with well-defined results. To excite the sciatic nerve it was only necessary to bring it near the electrodes. But if the electrodes were close together, and the current not too strong, there would be at one or two centimetres distance from them no diffusion.

It is not simply a question of peripheric diffusion, for the diffusion from the periphery to the parts underneath cannot be avoided. The gray layer is so thin that it cannot be expected to limit excitation to that part alone, so that the current necessarily extends to and excites the subjacent white layers.

---

[1] Onimus : Bull. de la Soc. de Biol., 1867 and 1874, p. 379.

[2] Cited by Bochefontaine, Arch. de Phys., 1876, p. 171, et Bull. de la Soc. de Biol., 1875, p. 131.

This suggested another experiment.[1] In place of exciting the gray substance, that can be removed, and the white substance underneath excited.

Many writers have made experiments, but without accord in results.

Putnam,[2] on the one hand, has observed that in removing a bit of the gray cortex and exciting the subjacent white substance a stronger excitant was required to produce a movement. In replacing the bit which had been cut out, the currents were without effect; he therefore concludes that the gray substance itself is susceptible of excitation.

Carville and Duret had a similar experience respecting the necessity of a stronger current, where the gray substance had been removed.

Hermann[3] and Braun,[4] on the other hand, have obtained quite different results. Hermann shows that, after destroying the gray substance with chemical cauteries, a very feeble current sufficed to produce movements, and that in cutting away slices from the brain the effect was decided in proportion as the central regions were approached. In some cases, however, it was necessary to increase, in others to diminish the force of the current

To this, Braun has added the important fact, that if the white fibres beneath the point excited be cut, the excitation fails to produce the movement which occurred before the section of the white fibres. The section does not prevent the receiving of currents, though in exciting the surface the corpora striata are not excited, but only the subjacent white substance or the gray substance itself.

---

[1] Upon this subject see Vulpian's lesson, June 29th, 1876 ; in Journal l'Ecole de médicine ; Carville and Duret, Arch. de Phys., 1875 ; Bourdon-Sanderson, Proceed. Roy. Soc., June, 1874, xxii., p. 338 ; Furrier, Functions of the Brain, p. 218.

[2] Boston Med. and Surg. Journal, July, 1874.

[3] Ueber elektrische Reizversuche an der Grosshirnrinde, Pflüger's Archiv., t. x., p. 77.

[4] Eckhard's Beiträge, etc., 1874, t. vii., p. 127. Beiträge zur Frage über die elektrische Erregbarkeit des Grosshirns.

When the gray substance which has been removed from above the white is replaced and electrized, it will be observed to have become inert, resulting from the section.

Other well-attested facts prove also that in exciting the cerebral periphery the corpora striata are not affected. Carville and Duret observed, in experimenting with a dog, that very strong electric currents would produce no muscular movements; autopsy brought to light a considerable lesion of the centrum ovale, thus interrupting the physiological, though not the physical continuity between the corpora striata and cortex cerebri. The corpora striata were sound. The excitation then was not diffused beyond the border of the corpus striatum.[1] Ferrier remarks that excitation of the corpora striata, or of the peduncles, gives quite different results from excitation of the cortex cerebri.[2]

Franck and Pitres, in the remarkable researches which they have undertaken upon the functions of the cerebral hemispheres, have often noticed that electrization of the corpora striata, carefully avoiding the white fibres which penetrate the nuclei of the gray substance, is absolutely without results.[3]

The general conclusion then is : in exciting the cortex cerebri, there are currents of diffusion to the periphery and towards the centre ; but these currents are insufficient to excite either the entire periphery or the subjacent central ganglia.

Respecting the greater or less degree of excitability of the white substance, when the gray substance is avoided or destroyed, there are great differences of opinion. In closely examining the facts, however, that discord is found to result from a difference of conditions.

---

[1] Carville and Duret : Notice of a pathological Lesion of the Centrum Ovale in a Dog. Arch. de physiol., 1875, p. 136.

[2] Functions of the Brain, French trans., 1875, p. 258 et suiv.

[3] A portion of the researches of Franck and Pitres has been communicated to the Soc. de Biol., Nov. and Dec., 1877. See Gaz. méd. of Jan. 3d, 1877. But many of the facts here given are unpublished, and we are indebted for them to our good friend, Fr. Franck.

In one of the experiments which I made with Bochefontaine, in the laboratory of Vulpian, the following facts were
observed :

A dog was chloralized and the sigmoid gyrus exposed.
In exciting the anterior part by a current of variable intensity, it was found that in order to provoke a movement it
required an electric current (induced, continuous) corre-
ponding in strength to No. 12 upon the indicator of Dubois-Reymond.[1]

After cutting away the gray substance and exciting the
white substance immediately underneath (before the occurrence of congestion caused by the cut), a very feeble current, scarcely sensible to the tongue—23, sufficed to provoke
motion.

After the lapse of an hour, excitability had greatly diminished, 11 being required to produce motion, and the excitability rapidly vanished.

This was followed by exposing the right hemisphere to
12, the gray substance did not respond to the excitation,
though upon removing the gray substance, the white substance responded to 12.

This would allow the conclusion that the white substance
is more excitable than the gray.

Still, as Franck has, in a large number of experiments,
uniformly obtained a great diminution of excitability after
removal of the gray cortex, and as my experiment just
given is very clear, I see no possible explanation of the dis-
agreement except in the difference of experimental conditions. The dogs upon which I experimented were chloralized, whilst Franck's were neither anæsthetized nor under
influence of curare. It seemed as though the chloral had
paralyzed the gray cortex, thus interposing an inert tissue
between the electric excitation and the white fibres which
alone were susceptible.

---

[1] That index, though very imperfect, is perhaps all that is necessary for physiological uses. With an ordinary Gremet pile, o indicates very strong, 10
middling, 20 very feeble, 30 perceptible only to the galvanometer.

These various experiments seem to prove that it is the gray substance which is really excited.

To these facts are associated other phenomena especially pertaining to the cortex cerebri ; they show that its reaction to excitation differs from the reaction of nerve-trunks. As I have had occasion to demonstrate elsewhere,[1] it appears that successive excitations do not accumulate in the nerve, whereas it is quite probable that they do in the receiving organs, whether muscular or sentient.

It was of interest to ascertain if the cortex cerebri acted in the same manner, and here is the result of experiment.

With a chloralized dog, excitation of the antero-superior portion of the sigmoid gyrus induced movements of the eyelids upon the same side and of the fore-legs of the opposite side. Currents were frequently repeated at a strength of 10, Dubois-Reymond indicator.

At 0° (maximum) with a single excitation (closed or open) there followed *no effect*.

A movement could be induced by making with the hand three or four tolerably rapid interruptions. In replacing the indicator at 10°, very frequent excitations produced movements both of eyelids and fore-legs, but, as has been noticed by Schiff, the movements were very retarded.

That retardation only signifies that the excitations are accumulated, and that they end in producing a result : the first excitations give no result, the last only (which includes the previous ones) does.

To explain this phenomenon it will perhaps be well to introduce one of my old tracings. It shows that frequently repeated excitations end by accumulation and produce no motion except when made with a certain frequency (fig. 10).

I have obtained phenomena which can be compared, by exciting the cerebral surface of a dog, and registering his movements.

After adapting an instrument to register the muscular

---

[1] Thèse inaug., Recherches sur la sensibilité. Paris, 1877.

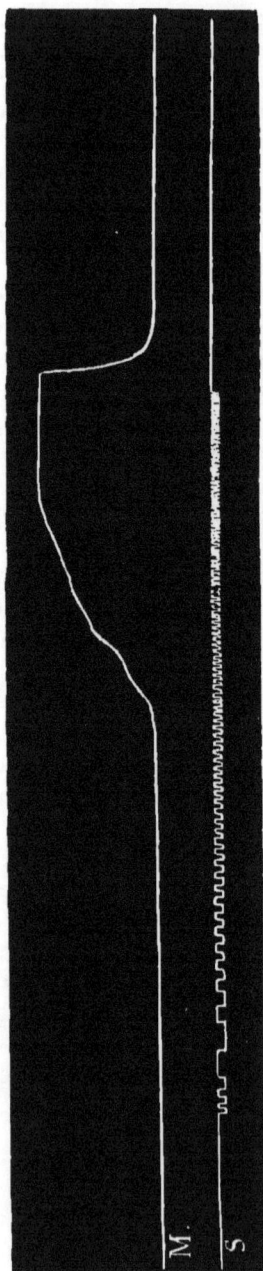

FIG. 10.—Latent addition in the muscle of a lobster. Induced current, augmented in frequency.

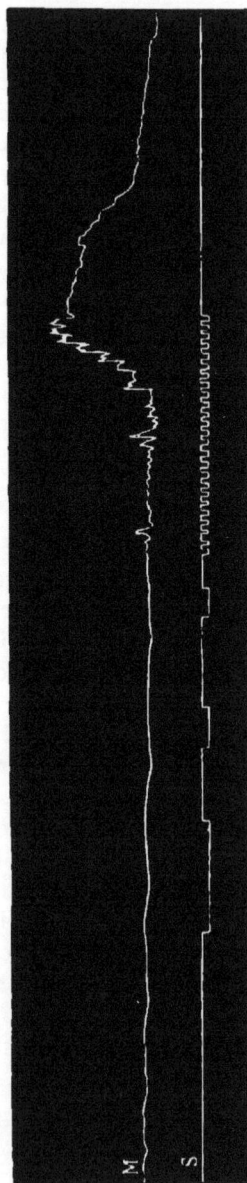

movements of the fore-legs, I observed that isolated excitations produced no effect, whereas closely succeeding
excitations produced a manifest tetanus (fig. 11).

In a similar experiment, MM. Franck and Pitres have
also observed this addition of excitation, even when they
were somewhat separated (fig. 12).

It cannot be supposed that we are here dealing with
latent accumulations either in the gray ganglionic substance or in the subjacent white fasciculi ; in fact, direct
excitation of these fasciculi, after removal of the gray cortex, gives no accumulation of excitations.

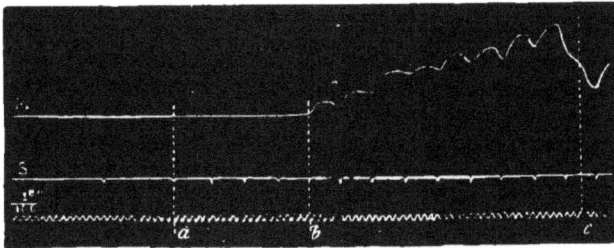

FIG. 12.—Addition of excitations in the substance of the gray cortex. From
a to *b*, excitations induced without effect. The line commences to wave at *b*,
and continues to increase. At the bottom, line of vibrations of the diapason.
(One hundred vibrations per second.)

As a result of all these facts it seems that the gray cortex
is directly excitable.

A number of other experiments also support this hypothesis. Pitres and Franck have shown that the interposi-
tion of the gray substance produces a retardation of $\frac{3}{200}$
in a second, a small figure of itself, but enormous considering the trifling distance of only some millimetres. This
indicates that the subjacent white fasciculi are not excited,
but rather the gray cortex, which responds very slowly
to the excitation.

It has been objected that abrasion of the gray cortex
does not prevent the action of electricity, the subjacent
white fasciculi alone being excited. But does that prove

that the cortex is not a centre? Not at all. Let us suppose, as Vulpian has said, that the abraded cortex is really a centre, it must have conductors which run to the deep part of the brain. These conductors are precisely the white fibres, and excitation of the conductors should give the same results as excitation of the centres from whence they start.

The reason adduced by Dupuy, that there were no centres in the cortex because chemical or mechanical excitation produced no reaction, is evidently insufficient ; indeed, the white substance which does not respond to chemical excitants evidently does respond to electricity ; so that, should any part of the nervous system may respond to chemical agents, that would be no proof that it would not respond to electricity.

We cannot dwell upon these facts, and will but add a resumé :

1st. Peripheric diffusion can be avoided, but diffusion from the gray to the white substance cannot be avoided.

2d. The white substance of the brain is certainly excitable, as Haller thought, and contrary to Flourens' opinion.

3d. The cortex cerebri is probably excitable by electricity, though it is nearly impossible to furnish direct proof of it.

4th. In the order of electric excitability of the nervous system, there may be admitted (though perhaps as yet somewhat hypothetically) : (a), the nerve terminations ; (b), nerve-trunks ; (c), central gray substance ; (d), white substance of the nerve-centres.

I will also call attention to an important fact hitherto imperfectly studied ; that is, the rapidity with which excitation disappears. It seems that the nervous centres are much more delicate than the nerve-trunks, and that exhaustion there ensues much more promptly. But, on the other hand, after a very short repose, excitability returns.

I will not dwell upon the facts which seem to prove that the cortex cerebri may give rise to epileptiform convul-

sions (Hughlings Jackson). Vulpian says that the manifestations in partial epilepsies always leave a considerable
doubt in the mind ; because, although a lesion may be distinctly visible in the cortex, still it is impossible to know
if the epilepsies really proceed from it. Partial epilepsies
often exist when there is no lesion appreciable either to
the naked eye or the microscope. The experimental epilepsy observed by Franck fully proves the excitability of
the gray substance.

It would be interesting to investigate how that excitement is modified by sanguiferous tension. The influence
of the brain upon sanguiferous tension has heretofore been
investigated ; but the influence of arterial pressure upon
the excito-motor power of the brain would be a curious
study.

We are here brought to a consideration of the method
in which the cortical region responds to electricity. Hitzig
has especially employed direct currents, which afford better
results than the induced, interrupted current chosen by
Ferrier. Of the two poles, the anode (positive) acts more
energetically than the cathode (negative), and this difference is more marked in proportion to the feebleness of
the current. Schiff has also made some interesting observations upon the same subject, though they have led to
a theory certainly erroneous.[1]

The constant current acts better than the induced ; and
of the induced currents the open one is best. This is because it lasts longer; and Schiff says that for the sensations a certain duration of excitement is necessary. Motor
responses, he says, are not instantaneous, but require the
$\frac{1}{260}$ of a second ; whereas, if a nerve was employed as the
conductor, the interval would be but the $\frac{1}{3000}$ of a second.
He holds that it consequently involves a reflex action.

Now this conclusion is not exact, because the cortical
substance is evidently concerned in these phenomena.

---

[1] Lezioni sopra il systemo nervoso encefalico, Firenze, 1874, et lo Sperimentale, xxxvii., p. 239, xxxviii., p. 241.

6

A very important problem to elucidate, not only on account of the method, but in a general physiological point of view, is the influence of various poisons on the cerebral excitability.

Hitzig in his first memoir[1] has shown that, in animals etherized or under the influence of morphine, the sigmoid gyrus is excitable. He concludes by saying: "When animals are profoundly etherized, though all traces of reflex action have disappeared, the electric excitability of the brain is partly retained, partly lost. On the contrary, with morphine, even in large doses, the excitability is not diminished."

Carville and Duret[2] in their experiments have employed chloral with good results, it giving an absolute insensibility, though preserving, with some reduction, the cerebral excitability.

Bochefontaine,[3] in a remarkable series of experiments, shows that with the use of curare, which completely paralyzes the voluntary muscles without affecting those of organic life, it can be proved that the brain preserves all its excitability. This is an interesting confirmation of the celebrated experiments of Claude Bernard, and it is seen that curare affects the nervous system only as it is connected with the muscles of animal life. Any one may satisfy himself that, as has been shown by Schiff, when an animal is in a profound state of anæsthesia, all motor reaction disappears, the bulb only exercising its functions. Upon this fact, together with those just mentioned, Schiff based his reasons for considering the movements succeeding the excitation of the convolutions as movements of reflex origin, a theory to which we shall recur. We only observe that, without further demonstration, it ought not to be said that chloral or ether affects only the reflexes. On the contrary, it very probably affects all the nerve-elements.

---

[1] Loc. cit., p. 401   [2] Bull. de la Soc. de Biol., 1874, p. 377.
[3] Arch. de physiol., 1876, p. 140.

When chloralization has been pushed to the extreme, the bulb is the only vestige of nervous life retained by the torpid animal, and by it are maintained the rhythmic movements of respiration; under these circumstances it can be demonstrated that galvanization of the brain is not without effect. The following experiment affords proof:

A small, lean cat was etherized and tied. Into the crural vein I injected forty-seven grains of chloral. The skull was opened and the crucial fissure laid bare. Excitation of that region provoked not the slightest movement of the legs, but it suspended the respiratory rhythm. In different regions of the cortex cerebri various points were found, the excitation of which immediately arrested respiration. Notwithstanding the enormous dose of chloral, the cat lived and respired with regularity for nearly four hours. At various times we verified that excitation of the brain and of the sciatic would arrest respiration. At the end of three hours sciatic excitation was without effect. A little later the same was true of cerebral excitation, and thereafter the strongest electric currents had no apparent effects.

We may hold, then, that the respiratory action is the last to fail; for when the excito-motor power of the limbs has disappeared, the action of the brain upon the bulb remains intact. It would seem that the excito-motor apparatus of respiration is the last to be paralyzed by poisons, both in the bulb and the cortex cerebri.

The effects of asphyxia on cerebral excitability have also been sought, but the results are not very concordant. Hitzig[1] found from asphyxia no action upon the excitability.

---

[1] Unters. über das Gehirn, Arch. für Anat., 1873, p. 404.

SEC. 2. THERMIC, ELECTRIC, AND CHEMICAL CONDITIONS OF THE CONVOLUTIONS.

Few experiments have been made upon the electric and thermic state of the convolutions. Schiff [1] and Caton [2] only have furnished some information upon the subject. Schiff, by means of thermo-electric apparatus, has proposed to measure the rise of temperature in the nerves and the nervous centres resulting from the influence of various excitations. I regret not being able to go into the details of these remarkable experiments ; I will give the conclusions only.

Sensible irritation of the peripheric nerves produced an increase of heat in the brain; excitation of the special senses, hearing, smelling, etc., had same result, so also with vivid impression, unexpected view of an object. Indeed, all mental activity expressed itself by augmented heat in the cerebral hemispheres.

Other authors (Broca, Voisin) have observed the temperature of the skull [3] and obtained similar results. All lively impressions or mental labor augmented the exterior heat of the skull, often it was confined to one side, generally the left (Broca).

It is probable that this difference in the external temperature corresponded to a difference of temperature in the deep parts of the hemispheres.

It might be supposed that thermic oscillations belonged

---

[1] Arch. de phys., t. iii., 1870, pp. 1, 198, and 451.

[2] Brit. Med. Jour., 28th Aug., 1875, p. 278. It is by an error that in the Revue des Sc. méd. the quotation from Caton is translated as though the experiments had been made by Ferrier.

[3] Lombard, Expériences sur l'influence du travail sur la température de la tête, analyse dans les Arch. de physiol., 1868, t. i., p. 670. Broca, Congrès de 1877. Voisin, Leçons sur les maladies mentales. France médic., 10 juillet, 1878. In course of publication.

to change of cardiac rhythm or to a difference in the local tension of the blood-vessels. Schiff rejects the first proposition, but is not so sure concerning the second.

Whatever hypothesis may be adopted concerning the cause of these thermic phenomena, the fact of itself is very important and accords with that which we know concerning the chemical activity of the encephalon. Experiments of some date back[1] have shown that cerebral activity increased the production of carbonic acid, urea, and probably also cholesterine. Consequently the nervous excitement which puts into activity the cerebral cells increases also the temperature and the chemical combustion.

There is a third phenomenon correlative to the chemical phenomena, that is the variation of the electric condition of the brain. It has been observed by Caton and described in a short note. In the normal state, there is an electric current (positive) which goes from the cortex cerebri to the white substance (cut), or into which a galvanometric needle (negative) has been plunged.

Those points in the cortex cerebri where electrization induces movements of the head and neck, and which in repose are positive, Caton has observed to become negative when connected with the white section, after sensorial excitations, especially after excitation of the retina. The current changes direction and develops an absolutely negative variation as in a nerve which is excited and the muscle which contracts. I have repeated this experiment upon a chloralized dog with Lippmann's electrometer which gives such precise indications, and I have been but partially able to verify these statements. It is true that if an electrode be placed at the surface of the convolutions and another to the deep parts, there will be found to exist an electro-motor power, less than that in the muscle and considerably more than that in the skin and fibrous tissues. But in exciting the sciatic, I have been able to discover

---

[1] Flint, Journ. de l'Anat., t. i., p. 565. Byasson, thèse inaug., Paris.

no variation in the direction of the cerebral electro-motor current. Perhaps the chloral-poisoning has prevented the phenomenon. However it may be, the experiment remains subject to repetition, and it would be interesting to follow the electric, thermic, and chemical variations of the brain, under the influence of sensorial or sensitive excitants.

The theory of these phenomena is too complex to be discussed here. To my mind, the theory of Dubois-Reymond and Pflüger is perhaps less satisfactory than that of Hermann, who explains the electric variations by chemical combinations. It is probable that the electric conditions depend upon increased chemical combustions, corresponding to increased nervous activity. .

SEC. 3. CIRCULATION IN THE CONVOLUTIONS.

We will not pretend to treat of cerebral circulation in a complete manner; it is proper, however, to speak of it as connected with the convolutions which are supplied with a rich and highly contractile arterial network.

Very precise experiments demonstrate that this abundant circulation is necessary to maintain the life of the nerve-substance. By ligating the carotids and vertebral arteries, encephalic circulation is more or less completely arrested[1] and with it the phenomena of encephalic activity cease.

Generally the circulation is at first completely abolished ; after a little it becomes re-established, and the lives of rabbits, and especially of dogs, can be preserved after the four arteries supplying the encephalon have been tied.

Vulpian has employed a still more certain process for

---

[1] For the history of the question I refer to the work of M. Couty, **Influence de l'encéphale sur les muscles de la vie organique.** Arch. de phys., 1876, p. 673.

the purpose of suppressing the circulation, not only in the spinal cord, but also in the brain, which is to inject the arteries with water containing a pulverized substance, powder of lycopodium for example, which serves to obstruct the minute encephalic arteries.[1]

From this process, sensibility and voluntary motion disappear with surprising rapidity, the animal remains inert, the bulb alone retaining its power of function.

Vulpian observed that in the spinal cord the gray substance only was paralyzed, the white preserving its normal conductibility.

Resting upon these fundamental facts, Couty has studied the effects of cerebral anæmia from arterial obstruction, and he first observed that this kind of anæmia did not affect the peripheric vessels, though galvanic excitation of the cortex cerebri does; consequently, galvanic excitation acts some other way than as a simple arterial constrictor.

This is worthy of note; for some authors, especially Brown-Séquard, without clearly defining his opinion upon the effect of electric excitation, maintain the idea of vascular constriction. It is not probable that he still holds that opinion as concerns galvanic excitation of the cortex cerebri. In carefully examining the cerebral surface when electricity is being applied, no vascular contraction is observed; on the contrary, there is dilatation, and capillaries before invisible, increase and become visible. I am aware, however, that the opposite effects have been seen, but I believe those to be complex effects, which attentive analysis can unravel. To the present there is nothing entirely positive and constant.

Couty has made another deduction from these experiments: that since the gray substance exercises no influence upon the vessels, and electricity does, therefore electricity does not directly influence the gray substance. The deduction is reasonable, but as it is always necessary in

---

[1] Leçons sur l'appareil vaso-moteur, t. ii., 1875, p. 118.

physiology to distrust indirect proofs, it would be better to make the direct experiment, and ascertain if the cortex cerebri of the animal whose encephalon is anæmied by the powdered injection is still excitable by electricity. This alone will permit a rigorous and indisputable conclusion.

Be this as it may, so far as concerns cerebral function, we see that complete anæmia rapidly suspends sensibility and voluntary motion.

We will now see what modifications belong to normal cerebral circulation.

Attention is at first called to the difficulties of experimentation. If the brain of an animal is exposed, contact with the air will excite or paralyze the vessels, and generally produce intense congestion. If it be shielded with a glass cover, the exuded blood and fluids prevent seeing what goes on; and besides, what could be concluded, for might not the consecutive encephalitis be a source of error difficult to eliminate? Moreover, to prevent voluntary movements of the animal, which, if violent, would suddenly change the venous pressure, chloral, morphine, or ether would be necessary, and consequently no conclusion of a certain nature could be arrived at.

Thus we have recourse only to indirect proofs or to inconclusive experiments. Physiological knowledge upon the subject is as yet very vague.

Some things, however, are precisely known. Long since, Claude Bernard [1] observed that, after section of the great sympathetic, the hemisphere of that side was warmer than the other. In this respect, the arteries of the brain acted in a similar manner to those of the eye, the ear, and the face. Vulpian [2] repeated these experiments by exciting instead of cutting the sympathetic nerve, and in various cases he has seen the vessels contract in a most notable manner.

Nothnagel states that excitation of the sciatic nerve

---

[1] Mém. de la Soc. de Biol., 1853, p. 94.

[2] Loc. cit., t. i., p. 109 ; t. ii., p. 120 et seq. In that work will be found a complete history of the question ; it is therefore needless to reproduce it here.

causes a reflex dilatation of the vessels of the pia mater. But that reflex in the encephalic capillaries has been thrown in doubt, and it does not appear to me that the matter can be considered as settled, although Regnard, in an interesting work,[1] states that he has seen peripheric excitations produce cerebral congestion (Expér. I. et III.).

As Vulpian remarks, excitation of the sciatic (after section of the great sympathetic) often causes contraction of the vessels of the ear, and the same phenomenon is probably produced in the vessels of the encephalon. It is possible, then, that the vaso-motor action upon the periphery of the vessels from the pia mater may induce either dilatation or contraction, according as the great sympathetic is intact or not.

In its pathological aspect, Brown-Séquard thinks that the reflex contraction of the vessels of the pia mater might ·produce cerebral excitation and consecutive epilepsy by anæmia. This theory seems controverted by that which we have just remarked, concerning the immediate paralysis of the excitability of the brain by anæmia. Besides, Ferrier has observed that, accompanying an epileptic attack produced by galvanic excitation of the cortex cerebri of an animal, the surface of the cortex was congested and not anæmied. Vulpian has several times observed the same.

Many authors have dwelt upon the relation of sleep to encephalic circulation. It has been supposed that sleep was due to cerebral congestion. This was contested by Durham, who made the first regular experiments upon the subject.[2]

Removing a round bit of the skull, he laid bare the dura mater and examined the brain. From his experiments he concluded, that during normal sleep, the brain is anæmied and that upon waking the brain became congested, so that anæmia was either the cause or the effect of sleep.

Durham's experiments are, to be sure, few and open to

---

[1] Thèse de Strasbourg, 1868.
[2] Physiology of Sleep. Guy's Hosp. Reps., 1860, p. 149 et seq.

criticism. I would remark the same respecting observations made upon those ill or wounded.[1]

Regnard's experiments upon chloroformed animals prove nothing for the normal state, at least the demonstration is insufficient.[2]

Flemming[3] has made some researches which seem to confirm the theory that sleep results from anæmia. Strong pressure upon the carotids produces a passing hyperæsthesia, characterized by vertigo, ringing in the ears, hyperideation, analogous to that in sleep, finally sleep—sleep and anæsthesia. But what relation is there between these phenomena and real sleep?

Brown-Séquard has also defended the theory of cerebral anæmia as the cause of sleep, and to a certain degree he assimilates normal sleep to a light attack of epilepsy. He produces an ingenious likeness, but a phenomenon of which the cause is not known cannot be explained by a phenomenon yet more mysterious.[4] At all events, the theory of anæmia, supported by the facts of Durham and Flemming, carry no more conviction than other facts which would seem to prove the contrary.

The direct experiments undertaken for the purpose of judging the condition of the cerebral surface have never given uniform results.

Neither Regnard, Langlet, Durham, Hammond nor J. Cappie[5] have ever been able to precisely describe, by these methods, the condition of cerebral circulation during sleep.

Gubler[6] thought that observation of the pupils would give useful indications. In fact, it is admitted, and many

---

[1] Krauss, Gaz. hebd., 1854.--Brown, Am. Jour. Med. Sc., 1861, etc.

[2] See the excellent thesis of Langlet, Etude critique sur quelques points de la physiologie du sommeil, th. inaug., Paris, 1872.

[3] Anæsthesia by Compression of the Carotids, Bull. gén. de thérapeut.), t. xlix., p. 37).

[4] For an exposé of these opinions by Brown-Séquard, consult an analysis which he has given of Kussmaul and Tenner. Jour. de Phys., t. i., 1858, p. 201.

[5] The Causation of Sleep, Edinb., 1872.

[6] Gaz. des hôp., 1858.

facts seem to prove it, that the encephalic circulation and that of the iris are allied, congestion of the first always coinciding with irido-choroidian congestion. Now when the iris is congested, the pupil is contracted, consequently contraction of the pupil is a sign of cerebral congestion.

In normal sleep there certainly is almost always contraction of the pupil; but, as I have before said, should we not exercise prudence in accepting indirect proofs, requiring a series of reasoning which may be excellent in appearance, but perhaps in reality erroneous?

To affirm that a thing exists, it must have been seen, and unfortunately it has not yet been seen in normal sleep, uncomplicated by pathology or experiment, whether the brain was congested or anæmied. The question, then, is in dispute and the protocol incomplete.

Thanks to registering instruments, science has lately been enriched by some valuable knowledge relative to cerebral circulation.

Some time since, Magendie,[1] Bourgougnon,[2] and my father [3] studied the movements of the brain and the oscillations of the cephalo-rachidian fluid.[4]

These labors demonstrated that the encephalon becomes swollen during violent respiratory efforts, and that cerebral movements depend in part upon the cardiac impulse and partly upon the respiratory rhythm. In the hands of Salathé, Franck,[5] Mosso and Giaccomini,[6] the graphic method has given very remarkable results, which confirm and complete the opinions of previous observers.

Although these experiments could not, as was realized

---

[1] Jour. de la phys., t. vi., et t. vii., 1825.

[2] Th. inaug., Paris, 1835.

[3] Anat. méd. chir., 1st ed., 1857.

[4] For the bibliography see the very complete thesis of our friend Dr. Salathé, Recherches sur les mouvements du cerveau, Paris, 1877. The major part of the experiments there mentioned may be found in the Comptes rendus du laboratoire de M. Marey, for 1876.

[5] Researches upon Expansion of the Brain. Jour. de l'Anat., 1877, p. 267.

[6] Comptes rendus de l'Acad. des Sc., 3d Jan., 1877.

by Franck[1] and Salathé[2] serve to settle the question respecting the circulatory cause of sleep, nevertheless it gave positive facts relative to arterial tension in the brain.

The following are the principal facts thus brought forward :—

1st. In repose, in absence of all effort, movements of the brain do not correspond to the respiratory rhythm ; only to the arterial rhythm.

2d. Each systole of the heart increases the volume of the brain, a kind of congestion.

This fact respecting the brain is analogous to that which Piégu, Mosso, and especially Franck have observed in various other parts of the body, only on account of the enormous vascularity of the cortex cerebri the phenomenon is there much more emphasized.

3d. Inspiratory and expiratory efforts greatly change the movements and volume of the brain. Expiration and effort augment its volume, inspiration greatly diminishes it. Compression upon the veins of the neck increases it.

4th. The cephalo-rachidian liquid is the moderator or safety-valve which protects the distended cerebral pulp from pressure against the skull-walls (A. Richet).

We reconsider these facts first, because they are the only positive ones which we possess relative to cerebral circulation, and also that cerebral circulation is in reality the circulation of the convolutions. The white matter has little vascularity, and the central gray masses have a volume greatly inferior to the cortex cerebri. It may be said that about half of the blood sent to the encephalon is distributed to the cortex.

The convolutions, then, do not always contain the same quantity of blood ; with each action of the heart it varies, but this variation produces no functional disturbance.

The knowledge furnished by pathology relative to cerebral congestion or anæmia is of minor value (heart disease, maladies of the great sympathetic, migraine, disorders of sleep, plethora, etc.), and we will not here discuss it.

---

[1] Loc. cit., p. 285.   [2] Loc. cit., p. 45.

CHAPTER II.

FUNCTIONS OF THE CONVOLUTIONS.

Unfortunately we cannot treat of this vast question in its entirety. We will attempt, however, to bring out the positive points which the various methods of investigation have given to science.

Convolutions can be said to have three principal functions: motion, sensation, and intellection; we will examine them successively.

Sec. 1. FUNCTION OF MOTION.

A. METHODS OF INVESTIGATION.

Method by excitation.—We have before spoken of the methods of electrizing certain portions of the cortex cerebri. Some points only remain to be noticed.

a. The current must not be too strong.

b. The electrodes should be near together.

c. Care should be taken that the surface is not covered with blood.

Without all these precautions there is greater liability to diffusion, and therefore the conclusions cannot be so exact.

Furthermore, shaking of the brain, loss of blood, prolonged exposure of the convolutions to the air, should be avoided.

The results of different experiments cannot be considered as susceptible of comparison unless the experimental conditions have been the same. (Chloroform, chloral, morphine, etc.).

Electric excitation is an excellent process, but as it is subject to various grave objections it should be corrected by other methods.

Method by destruction.—Abrasion was first employed by Fritsch and Hitzig. Upon removing with the bistoury a thin bit of cerebral substance, paralyses follow. These have been especially studied by Carville and Duret.

The cerebral bit may be taken away with a scraper in place of a bistoury, or still better, may be cauterized with a hot iron. According to Carville and Duret all these processes give about the same results. These authors have shown that dogs so operated upon present certain paralyses, or rather—the loss of motion being incomplete—pareses.

A second method has been employed by Fournié,[1] Nothnagel,[2] and Beaunis.[3] This consists in the injection of a caustic liquid (chloride of zinc, perchloride of iron, chromic acid) serving to destroy the cerebral parts with which it comes in contact.

On account of the diffusion of liquids and consecutive inflammation, it is doubtful if this method gives good results, at least for the cortex, though respecting the corpora striata and the optici thalami, interstitial injections seem to have rendered Nothnagel some tolerably precise facts.

Along with interstitial injections should be placed Goltz's method of injecting compressed water into different points of the hemisphere. This process, which destroys the cortex upon a somewhat extended scale, does not seem adapted to determining the motor regions of the convolutions.

The third method we have before spoken of, it is that of Vulpian and Couty.

---

[1] Experimental Researches upon the Functions of the Brain. Paris, 1873.

[2] Virchow's Arch., t. lvii., lviii., lx., and lxii.; Experimentelle Untersuchungen über die Functionen des Gehirns.

[3] Traité de phys., 1876, note 5, p. 1101.

It consists in injecting powdered lycopodium and then examining the condition of the arteries in the brain, to ascertain precisely the cerebral regions which have been anæmied.

Experimental process, to a certain degree, makes known to us the general influence of the encephalon, but for precisely localizing a phenomenon as originating in a certain convolution or in a specified region, it is doubtful if this method is sufficient; but it may always serve as a means for correction.

In short, the method of superficial destruction, either by bistoury or red-hot iron, and that of electric excitation are those which have given and will give the greatest services. They have their inconveniences, but we know nothing better.

B. ACTION OF THE CONVOLUTIONS UPON THE MUSCLES OF ANIMAL LIFE.

When, as in the experiments of Fritsch and Hitzig, the superficial part of a dog's encephalon is excited by a moderate electric current that half of the body opposite to the hemisphere excited will exhibit movements, varying according to the points excited and differing somewhat also as the experimental conditions may differ.

The general character of movements thus induced differ considerably from those produced by excitation of the nerves or muscles. They are combined movements, limited to a group, or rather to a muscular function. They appear as if destined to some use. Besides this, they are much slower and more feeble than movements occasioned by direct nerve or muscular excitation.

The exact limitation of the action from motor-centres interests the physiologist less perhaps than the medical practitioner; much, however, has been written on the question.

The following cut gives an idea of the centres found by

Fritsch and Hitzig, and the explanatory text beneath sufficiently conveys their opinion.

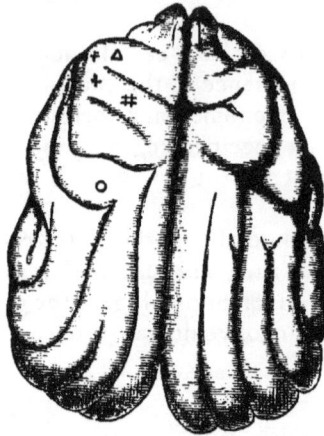

FIG. 13.—A dog's brain, serving to explain the researches of Fritsch and Hitzig.

Triangle, motor-centres of muscles of the neck. Cross and dot, centre of extensors and adductors of fore-leg. Cross, centre of flexion and rotation of fore-leg. Quadruple cross, centre of hind-leg. Circle, centre of facial nerve (?).

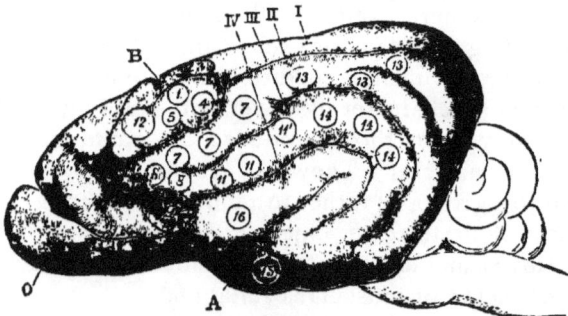

FIG. 14.—Right hemisphere of dog's brain (after Ferrier).

A, Fissure of Sylvius. B, Crucial furrow. O, Olfactive bulb. I., II., III., and IV., first, second, third, and fourth convolutions.

The first result acquired by Fritsch and Hitzig, and confirmed by Ferrier, Carville, Duret, and many others, is that only the region surrounding the crucial furrow is motor.

To this rule, however, there are some exceptions. According to Hitzig, movements of the muscles of the neck could be excited at the point indicated by the triangle, though these movements were not produced with constancy. Ferrier, in his numerous and interesting experiments, differs somewhat from Hitzig. The preceding cut (fig. 14) illustrates the localizations given by Ferrier:

1. The hind leg is advanced as for walking.
3. Undulatory motion of the tail.
4. Retraction and adduction of fore-leg.
5. Elevation and forward movement of shoulders.
7. Movements of the eyes.
8. Retraction of angle of the mouth.
9. Opening of the mouth and barking.
11. Retraction of angle of the mouth.
12. Opening of the eyes; head turns to the opposite side.
13. Eyes turn to the opposite side.
14. Ear becomes erect.
15. Twisting of nose to the same side (?).

These experiments have been repeated upon jackals and cats, and even upon animals where the convolutions are scarcely developed. They have no great interest though, especially as the exact localization of these excitations, as related to the resulting movements, is in no way certain.

Experiment has been made upon the monkey by Hitzig and frequently repeated by Ferrier. In the following cut Ferrier has represented the various excitable points of the cortex cerebri of the monkey :—

1. Leg advances as in walking.
2. Complex movements of thigh, leg, and foot.
3. Movement of tail.
4. Retraction and adduction of arm.
5. Forward extension of arm and hand.
6. Supination and flexion of fore-arm.

7

7. Zygomatic action, drawing the mouth backwards and upwards.

8. Elevation of ala of the nose and upper lip.

9 and 10. Opening of mouth with protrusion (9) and retraction (10) of the tongue.

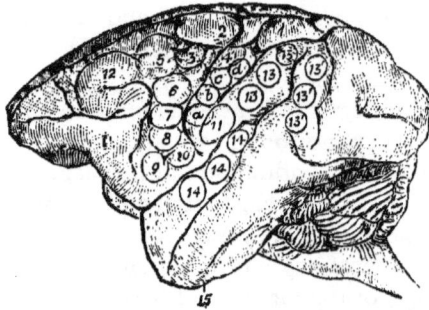

FIG. 15.—Lateral hemisphere of the monkey (after Ferrier).

12. Eyes open and turn to opposite side, pupils dilate.

13. Eyes turned to opposite side, raised (13), lowered (13′). Pupils contracted.

14. Pricking of opposite ear, pupils dilated, head and eyes turned to opposite side.

On account of the similarity between the monkey's brain and that of man, experiments made upon the monkey may serve to determine the motor-centres in man, and Ferrier has indicated upon the preceding schematic figure how his experiments may be applied to the human subject.[1]

In the monkey, excitations about the fissure of Rolando give the same results as excitations of the sigmoid gyrus in the dog. Elsewhere, at the periphery of the occipital, or even the frontal lobes, no movement follows.

The question is less simple, however, than one would at first believe, and is not settled so but that there still remain uncertainties.

---

[1] Functions of the Brain, French trans. by Duret, 1878, p. 222. English Edition, p. 304.

Thus Hitzig,[1] in other experiments, seemed inclined to believe that the anterior portion of the sigmoid gyrus (that which in his design responds nearly to the letter N, and the sign A), is not excitable by electricity, and may be removed without producing paralysis. The posterior part of the gyrus (that corresponding to E, and 1, 5, and

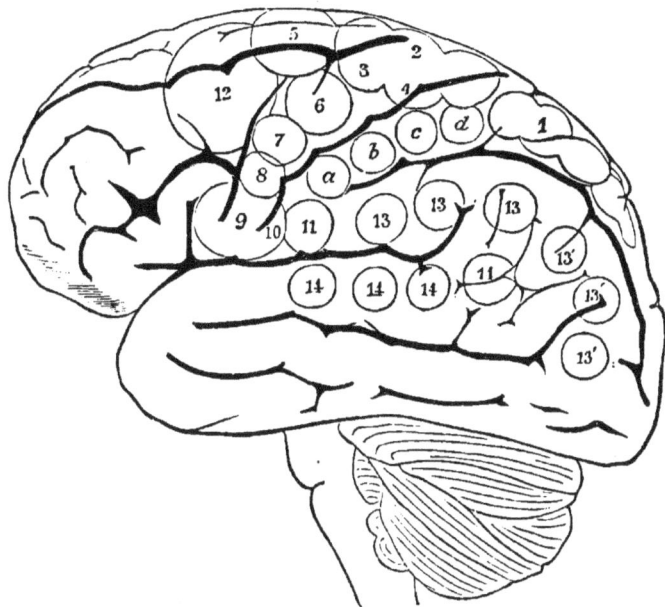

FIG. 16.—Lateral view of human brain.   (Letters same as for preceding cut.)

4 of cut 14), on the contrary, cannot be destroyed without consequent paralysis. With this paralysis there is a series of complex phenomena, difficult to disentangle, and which Schiff calls ataxia, and which Hitzig considers as significant of the abolition of muscular sensibility ; besides the phenomena of paralysis (absence of energy, will), there are also phenomena of special anæsthesia, particularly muscular anæsthesia.

---

[1] Neue Untersuchungen, Arch. für Anat., 1874, p. 432.

Marcacci [1] has recently observed that the excitable zone of the sheep's brain is chiefly in front of the crucial fissure, where there are four distinct centres, one for movements of the fore-legs, one for the neck, one for the face and tongue, and one for the movements of the jaw. No distinct centre is found for movements of the hind legs.

According to Ferrier, the anterior part of the sigmoid gyrus produces movements either in the head, eyes, or neck. I will add, that in experiments made with Bochefontaine, we have confirmed this statement, and we have also observed that with quite moderate currents, movements were induced either in the eyes or eyelids.

We have seen electrization of the anterior part of the sigmoid gyrus of a chloralized dog provoke contraction of the orbicularis palpebrare of the same side, and this where the influence of the dura mater was out of the question, as it was cut to a considerable distance, and upon being excited did not induce the same reflex.

Another fact seems to prove that there is not in the manifestation of any given movement the regularity sought for. Indeed, with the same dog, the electric excitation at the same point back of the crucial furrow, without changing position of the electrodes, we have frequently seen were dependent upon the strength of the current, movement of the fore-leg, movement of the hind-leg (with very feeble movement of the fore-leg), and a very forcible movement of both fore and hind-legs.

Consequently, absolute, inflexible localization of the motor-zones is all but impossible. There are zones which encroach upon each other, but none of these zones have limits of determined, rigorous constancy. The best proof of this is the difference existing among authors.

If I were to venture an opinion on the subject, I should say that so far as concerns details, the point is of small importance. It is of no special importance to know if there

---

[1] Arturio Marcacci, in Rendiconto delle ricerche sperimentali eseguite nel gabinetto di fisiologia della R. Universita di Siena. Milan, 1876.

is a centre for the ear, and exactly how many millimetres it is distant from the centre for the pupil. That which is important is to know if there certainly are centres for certain determined movements.

It is already proved (as respects the dog) that the posterior part of the crucial furrow is the eminently excitable region; above for the hind legs, below for the fore-legs.

Concerning centres for other muscular movements, I refer to Ferrier's work. Their existence is more questionable and additional investigation is evidently necessary.

We need not here consider the excitation of the surface of the cerebellum from which Ferrier observed movements of the eyes, we would only remark that the phenomena of diffusion towards the bulb ought to be more pronounced than that following the excitation of the hemispheres.

Electricity not having given entirely indisputable results,[1] the adjunct of another process has naturally been sought ; that of abrasion.

According to Carville and Duret, this operation gives the following results :—

1st. The paralysis is limited to a well-defined group of muscles.

2d. It is intermittent.

3d. It disappears at the end of five or six days.

Generally the paralysis is not complete, it is a sort of lameness, so that the dog operated upon cannot bring the foot to place and so walks upon the back of it. Goltz remarked that the dogs could not give the paw, but that in combined movements they contracted the muscles very well, when such movements were induced by the reflexes.

Schiff has dwelt upon the exhibition of ataxia and consecutive movements resulting from ablations of the cortex, which he compares to the phenomena arising from section of the posterior columns of the spinal cord.[2]

---

[1] See that before said, p. 65 et seq.

[2] Account of similar phenomena will be found in the memoirs of Goltz, Ueber die Verrichtungen des Grosshirns, Pflüger's Arch., t. xiii., and of Hitzig, Neue Folge, etc., Arch. für Anat., 1876, p. 692.

Albertoni and Michieli [1] from their labors have given the following results :

A. In rare cases there is no paralysis, though the operation has been exactly at the sigmoid gyrus.

B. The effects are more pronounced, defined, and durable with dogs than with rabbits.

C. Paresis in dogs diminishes upon the next day after operation, with rabbits it has by that time entirely disappeared, and at the end of four or five days it disappears in the dog.

It seems certain that in some cases there are more or less extensive destructions of the motor region of a cerebral hemisphere without consequent paralysis. Upon this subject may be recalled an experiment of Renzi's, cited in their remarkable work by Lussana and Lemoigne.[2] It is stated that the power of standing was unimpaired, but the body was inclined to the right side.

The instance published by Bochefontaine [3] is still more significant. " Vulpian, in one of his lectures of the course of 1875, repeated upon several dogs the experiments of Hitzig. In one the operation terminated, the wound was stitched up, and there was found complete absence of paralysis. The animal was kept ; some days after he was bitten by a another dog, the sutures were torn out, and a portion of the brain, in appearance like a reddish pulp, protruded through the opening in the skull. The wound healed without trouble, and for two months, during which the dog was kept under observation, there was no paralysis."

The results, then, of the method by abrasion are :

A. With a great majority of dogs, the ablation of the convolutions of the gyrus produces paralysis.

B. These paralyses are transitory.

C. With a small number of dogs there are no paralyses.

[1] Lo Sperimentale, Feb., 1876.
[2] Des centres moteurs encéphaliques, 1877, Arch. de phys., p. 121.
[3] Bull. de la Soc. de Biol., 1875, p. 387.

Now as to the phenomena resulting from pathological causes.

1st. There are a certain number of instances (very few, however) in which an entire cerebral hemisphere has been destroyed without effecting paralysis. Several cases may be cited. In a few words I will condense the following one borrowed from Prof. Porta and recounted by Lussana and Lemoigne.[1]

"*A young woman had an abscess upon the forehead . . . by use of a sound, destruction of the corresponding lobe could be recognized.*

During the last three days of her life, to the last moments, when convulsions and coma terminated life, the subject retained her mental, sensorial, and motor faculties as entire as though the brain had been uninjured. Post-mortem revealed the right hemisphere entirely suppurated, that is, converted into a yellowish-gray puriform substance . . . entirely disorganized and destroyed."

As it is not my intention to enter pathology, except so far as it will serve physiology, I will not recite other analogous cases which may be found (few, to be sure, are well authenticated) in the old archives of surgery or in the bulletins of the Anatomical Society,[2] etc. As Charcot says, very many of the reports of cases are without value, but it would be difficult to deny the truth of all.

Thus, with the human subject, as with the dog, there are exceptions to the general and well-established law, that destruction of certain parts of the cortex results in paralysis of certain muscles.

2d. The characters of these paralyses are the same as those of experimental paralysis, and I can do no better than to repeat the words of Charcot.[3] "*There are hemiple-*

---

[1] Loc. cit., p. 122.

[2] See also the memoir by Brown-Séquard in les Arch. de physiol., 1877, p. 655.

[3] In the thesis inaug. of M. Landouzy, Paris, 1876, p. 56 ; Convulsions et paralyses liées aux méningo-encephalites.

gias which may be called cortical, in contradistinction to those called central. The cortical paralyses are limited, transient, and variable ; the central paralyses total, embracing the entire one-half of the body, and always presenting the same characters ; cortical paralysis is abnormal, partial, so that it may be a monoplegia or may include the surface only."

These curious modifications of cortical motor-innervation will be examined further on, in the theory of these phenomena.

Some questions are yet to be resolved, for which pathology furnishes valuable knowledge.

1st. Can lesion of a convolution or of a very limited region of the cortex cerebri of itself produce paralysis ?

We could cite a great number of cases,[1] but we will content ourselves with recalling the following very clear one :

Without loss of consciousness, a consumptive was affected with a sudden weakness of the left arm. The feebleness increased until death, which occurred on the fourth day after. The paralysis was much more pronounced in the fore-arm than in the arm, and especially in the muscles supplied by the radial nerve. No sensorial disturbance.

AUTOPSY.—*A tubercle, the size of a millet seed, was found imbedded in the cortex, surrounded by a zone, one centimetre in diameter, of red softening. The lesion was located upon the posterior border of the fissure of Rolando (ascending parietal convolution), five centimetres and a half from the upper border of the hemisphere. The brain was otherwise absolutely sound.*[2]

2d. Where are the pathological alterations of the cortex located which produce paralyses ?

---

[1] See Charcot and Pitres :—Contribution à l'étude des localisations dans l'écorce des hémisphères du cerveau ; in Revue mens. de méd. et chir., 1877, Nos. ii., iii.   Bourdon :—Rech. Clin. sur les centres moteurs des membres, in Bull. de l'Acad. de méd., 1875, 2d série, t. vi., No. 43.—Foville: Ann. méd. psych., t. xvi., Jan., 1877. Bull. de la Soc. anat., 1875, 1876, 1877, 1878. Bull. de la Soc. Biol., 1876, 1877, 1878.—Lépine, in Revue mens. de méd. et de chir., mai, 1877.—Ferrier, British Med. Jour., March and April, 1878.—Bull. de l'Acad. méd., 1877, observations of Lucas.—Championnière, Terrillon, et Proust, reported by Gosselin.

[2] Maurice Reynaud :—Bull. de la Soc. anat., 25 juillet, 1876.

Observations of cerebral pathology, to a certain degree confirm the doctrine of cortical localizations. In the last few years, a large number of cases have been adduced and a very exact representation of the present aspect of the question may be found in the formulated conclusions closing the Mémoire by Charcot and Pitres;[1] they are deduced from the observations contained in the work:—

a. The cortex cerebri is not functionally homogeneous: only one part is concerned in the regular exercise of voluntary motion. That part, which may be called the cortical motor-zone, includes the paracentral lobule, the frontal and parietal ascending convolutions, and perhaps also the foot of the frontal convolutions.

b. No cortical lesions, whatever their extent, situated outside the motor-zone, affect the power of motion.

c. On the other hand, destructive lesions, even very limited, which affect either directly or indirectly the motor-zone, necessarily entail disturbance of voluntary motion.

Pitres[2] has shown that lesions involving the centrum ovale are not manifested by motor disturbances unless they affect the *fasciculi* subjacent to the zone of cortical motor-centres (fronto-parietal fasciculi) ; if, however, they affect the præfrontal, occipital, or sphenoidal fasciculi, they produce no motor trouble. He holds that the fibres composing the fasciculi of the centrum ovale are conductors, a section of which prevents manifestations from the cortical centres, as absolutely as the cutting of a telegraphic wire interrupts the current for telegraphy and renders useless the galvanic battery.

Such, in a general way, is the topography of the cortical motor-zone. But would it not be possible to determine somewhat more precisely the motor-centre of such or such

---

[1] Loc. cit., p. 456.

[2] Recherches sur les lésions du centre ovale et des hémisphères cérébraux etudiées au point de vue des localisations cérébrales. Thèse, Paris, 1877.—See also Ballet, Gaz. méd., 1878, No. 2.

a limb or of the various muscular groups? Examination of facts has enabled Charcot and Pitres to say that the cortical motor-centres for the opposite limbs are situated in the paracentral lobule, and in the upper two-thirds of the ascending convolutions; and that centres for facial movements are situated in the lower third of the ascending convolutions, in the neighborhood of the fissure of Sylvius.

Though in fact this only concerns the lower part of the face, inasmuch as cerebral lesions give rise to a hemiplegia which is always limited to the lower portions of the face, the superior parts remaining free (orbicularis palpebrarum, superciliaris, frontalis), a symptomatic dissociation which gives us a right to seek a correlative anatomical dissociation.

It is probable that the centre for isolated movements of the upper extremities is located in the middle third of the ascending frontal convolution. The exact situation of the cortical motor-centres for the nape of the neck, the neck, the eyes, and the eyelids is not at present known.[1]

Respecting instances of united deviation of the head and eyes from hemispheric lesion, there is at present no absolutely satisfactory solution.

Is paralysis a necessary consequence of destruction of the motor-zone? Charcot. as will be seen, says yes. But such is not the opinion of all physiologists. Vulpian and Brown-Séquard cite some exceptions to the rule, and remark that, if there exist cortical centres, their suppression ought invariably to entail a loss of their function. Further on, this question will be examined as relates to supplementation.

The three questions which have been proposed may be answered thus:

1st. Cortical lesion alone can produce a permanent or a transitory paralysis.

[1] Notwithstanding an observation of Grasset's, Progrès méd., 27 mai, 1876, p. 431. See also Landouzy, Arch. gén. de méd., 1877, août.

2d. Cortical lesion produces paralysis only when it is seated in the motor-zone.

3d. All cortical lesions of the motor-zone produce paralysis (there are exceptions to this law).

An important discovery confirms these different results; that is, descending sclerosis from the brain; it has been shown that in certain lesions of the motor-zone there was sclerosis of the spinal cord, consequently that the cerebral periphery is intimately allied to certain fasciculi of the spinal cord and the bulb.[1]

The motor action of the cortex cerebri may manifest itself in many ways; the action may be normal or convulsive, and some writers have made interesting remarks upon that subject.

Fritsch and Hitzig,[2] in their work of 1870, report having seen electric excitations of the cortex cerebri provoke convulsive contractions. These convulsions at first confined themselves to the muscles responding to the cerebral region excited, then became more general and extended so as at length to become truly epileptiform.

Ferrier repeated these experiments and found that with animals under the influence of anæsthetics there were no epileptiform convulsions, whilst with those not chloralized they were easily provoked.[3]

Other experiments in this field have been made, especially by Albertoni,[4] who describes an epilepto-genetic zone, which appears to be the same as the motor-zone of Hitzig and Ferrier.

---

[1] Bouchard: Secondary degenerations of the spinal cord, Arch. de méd., 1866, p. 443, t. i.—Cotard: Study upon Partial Atrophy of the Brain, thèse, Paris, 1868, obs. iv.—Charcot: Localizations in Diseases of the Brain, 1875. Lépin: De la Localisation dans les maladies cérébrales, thèse d'agrégation, 1875, p. 53.—Pitres: Soc. de Biol., 21st Oct., 1876.—MacDonnel: Dub. Jour. Med. Sc., Nov., 1877, p. 451.—MacDonnel: Brit. Med. Jour., 14th July, 1877, p. 49.—Vulpian: Destruction of the Sigmoid Gyrus of a Dog, Arch. de phys., 1876.   [2] Loc. cit., p. 317.   [3] Loc. cit., p. 208.
[4] Rendiconto di esperienze fatte nello gabinetto di Sienna, 1876, 2d semestre.

Franck and Pitres have made experiments upon cortical epilepsy, and it is through their kindness that I am able to reproduce here some of their tracings (figs. 17, 18).

It will be seen that excitation of the intact cortical zone produces phenomena quite different from those resulting from excitation of the subjacent white fasciculi. From the first there is a primary tetanus provoked by the direct excitation, but this is followed by a very remarkable secondary tetanus, which is entirely absent in excitation of the white fasciculi. The physiological interest of that experiment is easy to comprehend.

According to Franck and Pitres, the centre of the gray substance which appears to be the point of departure may be removed during the provoked epileptic attack, and nevertheless the paroxysm will continue; precisely as if, immediately after transmitting a telegraphic dispatch, the first end of the wire were to be cut; that cutting would not destroy the dispatch which had gone on its way.

From the observations of these authors it may be concluded that cortical epilepsy results from the accumulation of excitations in the gray substance, which excite in the way of successive discharges.

In experiments mutually conducted by Bochefontaine and Viel, and recited in a well-written work by the latter,[1] Viel has succeeded in producing epileptic attacks by another process.

Injection of nitrate of silver between the dura mater and the brain produces menigo-encephalitis accompanied by convulsions. *"At the onset," says Viel, "there are only varied phenomena of ataxia. There is an uncertainty in the movements of the limbs, or of a single limb on the side opposite the lesion . . . when inflammation supervenes, the attacks become clearly epileptiform, with dilated pupils, striking together of the teeth, excessive salivation, a tonic and a clonic period."*

---

[1] Symptomatologie de la méningo-encéphalite.—Thèse inaugurale, Paris, 1878.

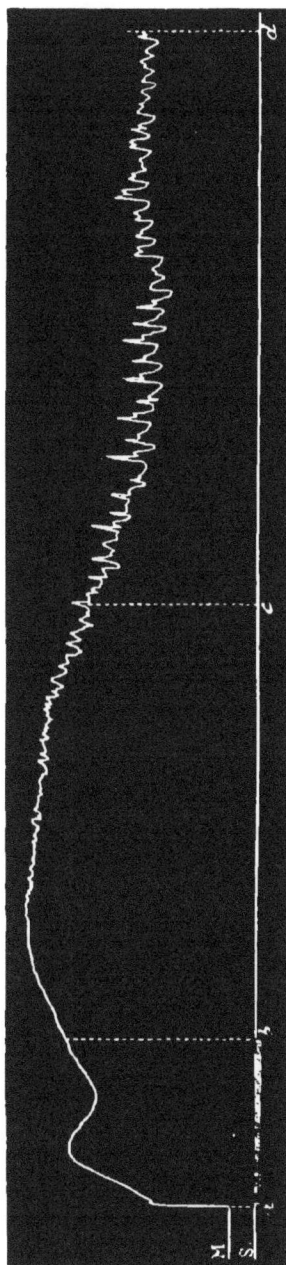

Fig. 17.—Prolonged epileptic attack provoked by cortical excitation, *a*, *b*, of short duration. From *b* to *c*, general tetanus (tonic period). From *c* to *d*, gradual dissociation of shocks (clonic period).

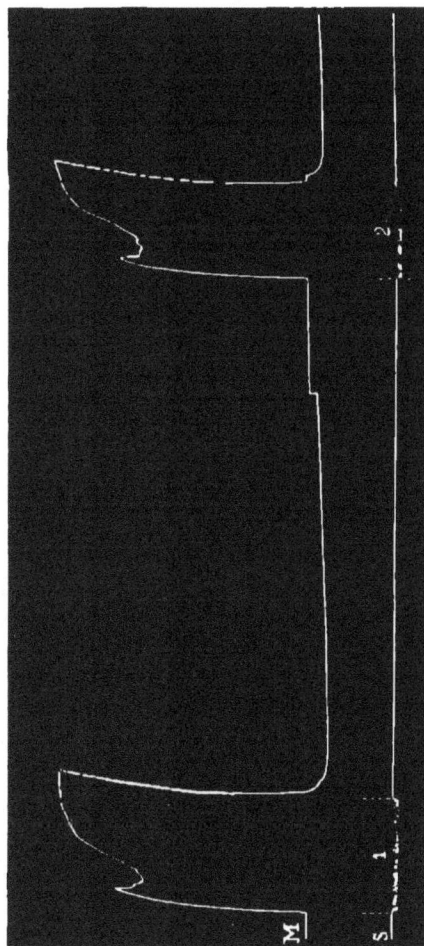

Fig. 18.—Tetanus provoked by very intense excitation of the white substance after ablation of the cortex which had been previously excited. No attack.

Convulsions, then, may be provoked by excitation of the cortex cerebri; but it does not appear that the cause of convulsions (electric), studied by Hitzig and Ferrier, is at all analogous to the cause of the convulsions (inflammatory) observed by Viel.

From the electricity there is at first a violent excitation of the sensibility, if not of conscious, at least of reflex sensibility. Known facts demonstrate that excitation of the convolutions which surround the sigmoid gyrus act with extreme energy upon the ganglionic centres of the brain (opto-striated bodies). It is possible that such excitation cumulates in the cerebral centres, and that these centres thus surcharged send successive discharges to the muscles. These are, of course, words which ill-cover our ignorance of the phenomenon.

In the experimental meningitis of Viel, inflammation replaced electric excitation, that is, substituted for the physical excitants one much more powerful. It is indeed the physiological excitant *par excellence*—inflammation—which so readily produces hyperkinesia and hyperæsthesia.

The result, in fact, is the same, and the phenomena, in foundation, are identical. Exaggerated excitation of the cerebral periphery produces convulsions, at first limited to a group of muscles, then to one side (the opposite) of the body, and which may, extending, involve all the muscles of the body. Is not this a very curious similarity to that which is observed in the study of reflex actions in the frog? In proportion to the increase of the excitant, the reflexes, which at first are localized, extend and become more and more general.

Medical practitioners have gathered important facts, which in a most formal manner confirm the results obtained by physiologists. Thus, more than twelve years ago, Hughlings Jackson[1] announced that convulsive move-

---

[1] His scattered memoirs have been gathered into a book: Clinical and Pathological Researches on the Nervous System.

ments of one side of the body depended upon a cortical irritation on the opposite side.

That form of epilepsy limited to one side of the body, or to one limb, or to the face, that kind of epilepsy which Charcot perhaps rightly calls Jacksonian, has been treated of in a remarkable work by Landouzy.[1] It is now known that a lesion of the cortex cerebri may produce either paralysis or convulsion (*akinesic, hyperkinesic*), according as the lesion is destructive or irritative. The convulsive centres coincide absolutely with the motor-centres, so that the two series of events are completed and confirmed by each other. In conclusion, I will observe that, besides convulsions, there is another modality of motor innervation—contraction. By an intense and profound excitant at the surface of the hemispheres, Ferrier has been able to produce contractions. In essence, however, the phenomenon is not different. Using the comparison already employed, the motor discharge, in place of being interrupted (clonic epilepsy), may be continuous (tonic contraction).

C. ACTION OF THE CONVOLUTIONS UPON THE MUSCLES OF ORGANIC LIFE.

The action of the convolutions upon the muscles of organic life was first studied by two of Vulpian's pupils, Bochefontaine and Lépine.[2] One year previous, Schiff thought that he observed accelerated action of the heart without change of blood-pressure.[3] At the same date, Külz[4] maintained that the encephalic centres had no action upon the saliva, so that the claims of Bochefontaine and

---

[1] Convulsions et paralysies liées aux méningo-encephalites. Thèse inaug., Paris, 1876.

[2] Bull. de la Soc. de Biol., 5 juin 1875, pp. 230–257.—Bochefontaine: Etude expérimentale de l'influence exercée par la faradisation de l'écorce grise sur quelques muscles de la vie organique.—Arch. de phys., 1876, p. 140.

[3] Arch. für experim. Path., 1874, 15th Dec., p. 178.

[4] Centralbl. für med. Wissens., juin, 1875.

Lépine to priority in this important discovery are beyond dispute. It should be observed, however, that Danilewski, in a communication to *La Société de médicine de Charkow*, Nov., 1874, remarked that excitation of the cortex cerebri produced a slight increase of blood-tension, with slowing of the pulse ; but the issue of his memoir is subsequent to the communication from Bochefontaine and Lépine.[1]

If a dog be curarized so as to paralyze voluntary movements, and yet retain organic life, and then the encephalon be laid bare, it will be seen that excitation of the gyrus will at once cause blood to appear in various spots on the wound. Hemorrhages which had ceased readily recommence ; there seems to be a considerable increase of arterial pressure.

This increased pressure is general and can be at once observed by adjusting a manometre to the carotid. The pressure augments very rapidly, and is often sufficient to expel a coagulum obstructing the glass canal placed in the artery. This excess of tension is also sometimes manifested by the increased volume of the brain, so that it protrudes through the skull as a cerebral hernia, an unfortunate circumstance for the experimental enterprise.[2]

Differences of pressure may be easily detected with the sphygmoscope, but it does not indicate the absolute pressure, and it is less sensitive than the kymographion. In the tracings collected by Bochefontaine it can be seen how marked the line of pressure is.

We here give the tracings from an experiment made, in company with Bochefontaine, in Vulpian's laboratory, upon curarized dogs, where the electric excitation and the arterial pressure in the carotid were inscribed simultaneously.

The principal results from the experiment, as may be seen from the tracing, are as follows:

---

[1] Experimentelle Beiträge zur Physiologie des Gehirns.—Arch. de Pflüger, t. xi., 1875, p. 128.

[2] Bochefontaine, loc. cit., p. 142.

A. Electric excitation, though brief, produces a considerable increase of blood-pressure which follows somewhat slowly, augmenting and persisting, even long after cessation of excitation.

B. There is a notable change in the cardiac rhythm; it at first becomes very frequent, afterwards much slower.

I will observe, moreover:

1st. These results arise from excitation of the anterior portion of the sigmoid gyrus.

2d. Although the excitation be far from intense, the animal becomes quickly exhausted.

3d. When exhaustion supervenes and the arterial pressure is lowered, electric excitation, even though powerful, has scarcely any effect.

Other experiments have shown to Bochefontaine :

1st. That sometimes slowing of the pulse is replaced by cardiac acceleration, the cause of which is as yet unexplained.

2d. That if the pneumogastrics have been cut, there is always, after cortical excitation, lowering of pressure and slowing of pulse.

3d. That excitation of the sigmoid gyrus may be compared to excitation of the sciatic, which raises the arterial tension and slows the pulse.

To this Danilewski has added an important fact, which is, that direct excitation of either the optic thalamus or the corpus striatum (except perhaps the caudated ganglion) does not produce a like elevation of pressure ; therefore, these effects are not due to diffusion of currents, but result from the excitation itself, either of the gray peripheric substance or of the white substance beneath.

Couty's method has led him by an entirely different road to analogous results.

First, he very justly excludes the results of older authors who tied the carotids and vertebrals ; for that ligation produces not alone anæmia of single convolutions, but of the entire mass of the encephalon.

8

Anæmia of the convolutions, produced by injection of pulverized substances into the carotids, first acts as an excitation to the convolutions. Indeed, all nerve-anæmia, before destroying, produces hyperkinesia of the nerve function. Now this intense excitation from anæmia does not increase cardiac tension, it slows the heart.

Continuing his experiments, Couty [1] has injected lycopodium into the carotids of curarized dogs in which the vagi were cut. Under these circumstances he has observed both cardiac acceleration and increase of arterial pressure. Bochefontaine has also found cardiac acceleration in dogs after the pneumogastric nerves were cut; it may therefore be admitted that, in the absence or destruction of the inhibitory nerves of the heart, encephalic excitation produces acceleration of the heart's action.

Respecting accrued arterial tension, I think it has not the value given to it by Couty, for Bochefontaine has seen excitation of the encephalon (in dogs deprived of the pneumogastrics) lower the pressure, and Couty does not sufficiently take into account [2] the arterial obstruction which evidently increases the blood-pressure: this, it seems to me, may solve the want of accord between the results of the two methods, when excitation is succeeded by paralysis of the cortex cerebri, the heart's action increases and tension diminishes; later on, the phenomena become complicated with spinal anæmia, this we do not enter into here.

This blood-pressure seems to arise partly from contraction of the peripheric arterioles and partly from excitation of the pneumogastric, which slow the heart and elevate the pressure. Excitation of the gyrus, however, produces other phenomena which I will hastily state according to the memoir of Bochefontaine.

In exciting points 1, 2, 3, 4 of Ferrier, a considerable hypersecretion of the submaxillary glands may be noticed and which can very well be compared to that produced by

---

[1] Loc. cit., p. 711.   [2] Loc. cit., 708.

lingual excitation.   Vulpian in his course has several times
repeated this experiment with very clear results.   When
the hemisphere is not excited, there is no flow, but when
the gyrus is electrized saliva escapes from the canula
placed in the ducts of Wharton.   The action is both direct
and crossed.

Hypersecretion of saliva has also been obtained from
points 11, 15, 10, and 17.

Movements have been observed, too, in the pupil, the
small and large intestines, the Fallopian tubes, the spleen,
the bladder, and a most abundant secretion of bile or of
pancreatic juice has been induced.   Upon these points I
refer to Bochefontaine's work.

But one thing is necessary to remark, which is, that all
these centres of movements, arterial or otherwise, are located in the same points from which motion of the limbs
is provoked ; that is, about the crucial sulcus.

The electric excitation must not extend to the dura mater ; for, as observed by Danilewski and Bochefontaine,
electric or even mechanical excitation of that membrane
provokes, by reflex action, salivary secretion, contraction
of the iris, etc.

Electrization of the sigmoid acts peculiarly upon respiration ; it is not correct to say that it accelerates it.   At first,
there is an irregularity and a certain acceleration, then a
pause, which often continues for half a minute.

I have seen this verified upon a cat profoundly chloralized, and have noted some interesting phenomena.   When
all other movements were suppressed, the action of the
bulb could be arrested by excitation of the brain.   In one
case, the arrest of respiration was so prolonged that death
would probably have ensued, had the thorax and body of
the animal not been vigorously electrized.   This agrees
with Vulpian's observation upon profoundly chloralized
dogs ; a violent peripheric excitation easily killed by suspended respiration, or by syncope.

That reflex arrest of respiration results also, though

with greater difficulty, from electrization of the sciatic nerve.

Be it as it may, the respiratory phenomena, observed in chloralized animals, entirely resemble that which Franck has noticed in rabbits upon excitation of the sensorial nerves of the face.[1]

Brown-Séquard has studied some of the phenomena to which excitation or destruction of the cortex cerebri give rise.[2] I have before said that the cerebral periphery was not insensible to excitation ; that in burning the surface of the convolutions with a hot iron, certain immediate phenomena could be observed ; congestion of the conjunctiva, closing of the eyelids, contraction of the pupil upon the side of the lesion.

For that eminent physiologist, these effects are identical with those obtained by section of the great cervical sympathetic. The degree of congestion appears, then, to be in proportion to the intensity of the excitation, and the extent of the cerebral surface cauterized. Brown-Séquard connects these facts with the phenomena of arrest, which he has elsewhere so well observed.

Similar facts, relating to the action of the Rolandic convolutions upon temperature, have been noted by other writers, by Heifler,[3] and especially by Eulenburg and Landois.[4] These authors state that excitation of the parietal convolutions produces a decrease of temperature ; whereas a destruction, by abrasion or liquid caustics, causes a rise of temperature.

The facts of Eulenburg and Landois are unfortunately very open to objection ; and several, particularly Vulpian[5] and Kussner,[6] have called them in question. Hitzig, cited

---

[1] Comptes rendus du laboratoire de M. Marey, 1876, p. 221.
[2] Arch. de physiol., 1875, p. 854.   [3] Wien. Med. Jahresber., 1875, p. 59.
[4] Centralbl. für med. Wiss., 1876, p. 260, et Virchow's Arch., 1876, t. xviii.
[5] Arch. de phys., loc. cit.
[6] Ueber vasomotorische Centren in der Grosshirnrinde des Kaninchens. Arch. für Psychiatrie, p. 432, 1878, t. viii.

by Kussner, also notes the absence of well-defined vaso-motor phenomena.

It seems, however, that a vaso-motor and thermal action, as yet little understood, is exercised by certain parts of the cortex cerebri upon the tissues, as exhibited in a dila-tation of the vessels (paralysis) and a contraction of them (excitation).

It is probable that the facts of experimental and clinical observations, following provoked or spontaneous cortical lesions, are due to this vaso-motor action. There may follow congestions of the lungs, stomach, intestines, etc. Of those dying with lesions of the brain, Ollivier has re-ported cases of subplural ecchymoses, pulmonary hemor-rhages, and renal congestions.[1] Under the same conditions, congestion of the liver, glycosuria, polyuria, and albumi-nuria, etc., have been observed.

These lesions and functional troubles appear to excite vaso-motor modifications of which the point of departure would be in the cortex cerebri. Lesions of the cortex seem to induce the vaso-motor disturbances daily noticed in clinic as a sequence of cerebral affections. We allude to those œdemas, erythemas, cutaneous congestions, often eventuating in acute illness, phases which rapidly super-vene upon the paralyzed side. To a certain degree they depend upon the same cause as do rise of temperature and diminution of tension in the vessels of the paralyzed side.

All these trophic troubles are acute in advent. There are other disturbances which appear later and which, physiologically, are of great importance: I refer to secon-dary lesions which, proceeding from the cortical lesion, go in direct line by way of the peduncle and protuberance to the lateral medullary column on the opposite side of the spinal cord. Secondary degenerations, true descending scleroses, may in all respects be compared to lesions pro-

---

[1] Ollivier, Bull. de la Soc. de Biol., 4th serie, p. 245. Modifications de la secretion urinaire après l'hemorrhagie cérébrale. Gaz. heb., 1875, Obs. I.

duced in the peripheric end of a nerve separated from its
trophic centre.   Thus there is authority to consider the
convolutions as veritable trophic centres of the nerve-fas-
ciculi which leave them.

Autopsies have demonstrated that descending scleroses
may result from lesions belonging exclusively to the cortex, the central ganglia being absolutely sound.[1]   These
scleroses are found not alone in man.   Seven months after
destroying the sigmoid gyrus in a dog, encephalitis having
followed, Vulpian[2] observed a descending atrophy of the
peduncle, isthmus, and the spinal cord.

It is under these circumstances that in the limbs opposite
to the side of cerebral lesion are seen slowly evolving trophic disturbances; scleroses, retractions and contractions,
muscular atrophy, arthritis, thickening of the subcuta-
neous cellular tissue, etc.   These troubles depend upon
sclerous lesions developed step by step along the spinal
column, and the results are similar to those which follow
lesions of the nerves; that is, the encephalic alterations
have produced trophic disturbances only by the intermediation of secondary neuroses.

It is also by the intermediation of these neuroses that
cortical lesions in infants entail arrest of development, mal-
formations and atrophies, so well described in the theses of
Turner (1856) and Cotard (1868).

If from these facts, the convolutions can be considered
as trophic centres of the nerves, as a system of nerve-pro-
jection, may it not be supposed, *a priori,* that there is a
relation between the mass of cerebral convolutions and the
peripheric nervous system ?

The problem is not yet solved, still there are a certain
number of facts which should be considered.[3]

[1] See Lépine : Thèse d'agrégat., 1875, p. 53, Paris, 1878.—R. Isartier :
Thèse inaug.—Des dégénérations secondaires de la moëlle épinière consécutive
aux lésions corticales du cerveau.   [2] Arch. de phys., 1876, p. 814.
[3] Luys, Bull. de la Soc. Biol., 8 juillet, 1876.—Féré : Bull. de la Soc. anat.,
mars, 1877.—Mossé : Bull. de la Soc. anat., février, 1878, etc.

When a long time has elapsed after the limbs have been destroyed or amputated, cerebral asymmetry of the Rolandic convolutions have been observed. In one instance atrophy was manifest, not only in the convolution, but also in the peduncle and bulb, the disease-evolution taking a direction inverse to that of descending sclerose.[1]

D. APHASIA.

Although the study of aphasia introduces pathology, still it seems proper to speak of it here, for the reason that physiology is the study of functions, and one of the functions of the convolutions in man being language, it is necessary to treat of it, even though in a summary manner. To tell the truth, our knowledge of aphasia has, of all the physiology of the convolutions, certainly the greatest precision and interest.

Aphasia is not a phase of motor paralysis. The muscles of the tongue, larynx, and velum palati retain their power of contraction, and their functions are unaltered. The trouble is a loss of ideo-motor coördination.

There is neither dementia nor paralysis, the defect is neither of intellection nor motion, but of the bond which unites the two, it being the effort of intellection which induces motion. Meynert was the first to describe the lesion of aphasia as a fracture of the *psycho-motor* centres. The term *psycho-motor* seems excellently applied to aphasia : it is not a motor centre, like the ventricle of the gray substance (fourth) which is the motor-centre of respiration, nor is it yet a psychic centre, since there is motor paralysis : it is a *psycho-motor* centre.

We say that intellection is intact ; but it would be erroneous to consider the intelligence of aphasiacs as unimpaired. One deprived of speech certainly has a lesion of intelligence, at least if intelligence embraces the totality of intellectual faculties. Besides, all medical practitioners

---

[1] Landouzy, Bull. de la Soc. anat., avril, 1877.

know that the majority of aphasiacs have childish ideas, weeping, etc.

Not only is speech lost in aphasia, but all other forms of language (imitation, drawing, writing, reading, singing) are more or less affected. Some aphasiacs can express neither negation nor affirmation by gesture of the head.

This is a very important fact, for it exhibits an intellectual function absolutely destroyed through lesion of a convolution, as if the convolution were the organ of that function.

Recurring to the question which we have before broached regarding the seat of the lesion, we can, thanks to Broca, return very precise answers:

1st. Lesion of one convolution, and even of a small area of its gray cortex, may of itself produce aphasia.

2d. That convolution is the posterior part (generally called foot) of the third left frontal convolution (*convolution of Broca*).

3d. Whenever the posterior portion of the third left frontal convolution is diseased, there is aphasia.

This last law, however, is not absolute. There are cases where the lesion has been in the island of Reil; others, in the right hemisphere; in other instances there have been lesions of the third left frontal without aphasia.

The exceptions may be classed thus:

A. Aphasia without lesion of third left frontal.

    a. With lesion of third right frontal.

    b. Without lesion of third right frontal.

B. Lesion of third left frontal without aphasia.

    a. With lesion of third right frontal.

    b. Without lesion of third right frontal.

I will not dwell upon facts which belong to medicine rather than physiology; it suffices to say that all these exceptions have been observed, but what do they prove?

If in one hundred cases of the same disease, the same lesion be found in ninety-nine, the exception being one, can it be said that there is no relation as cause and effect between the lesion and the disease ?

If in one hundred aphasiacs, ninety-nine have the third left frontal convolution destroyed, would it be justifiable to say that this is not the seat of the faculty of articulate language ?

For my part I think not, and I believe that the localizing of language in the foot of the third left convolution, or better, the convolution of Broca, is very firmly and amply established.

But we must fully consider the exceptions : they show to us that the convolution of Broca is not to language that which the retina is to vision, or the testicle to spermatogenesis. We cannot comprehend vision without a retina, or spermatogenesis without a testicle, but one can conceive that other parts of a hemisphere may replace those which generally preside over a function.

It is still easier to comprehend how the right hemisphere may replace the left : the same as there may be one left-handed person for one hundred right-handed ones, so there may also be one individual out of a hundred who speaks from the right hemisphere : that would be a left-handed speaker, as Broca happily expresses it.

We have seen that experimental cortical paralyses have a triple character : they are partial, inconstant, transitory. Such are also the traits often presented by aphasia. Unfortunately, this most interesting history of transitory aphasias, based upon autopsy, is not yet complete.

E. THEORIES OF MOTOR INNERVATION OF THE CONVOLU-
TIONS.

We have given the facts, and we will now see how they can be grouped so as to establish a theory and a system of unity.

Moreover, we will have occasion in this chapter to mention other facts which did not logically have place amongst the preceding ones.

All theories can be reduced to two principal ones.

A. There are motor-centres (Hitzig, Ferrier).

B. There are no motor-centres; there are but reflex (Schiff), or irritative actions (Brown-Séquard). It will be necessary also to discuss two hypotheses which may be called accessories, for the reason that they only seek to explain a part of the phenomena in order to support one or the other of the two theories, the supplemental hypothesis and the hypothesis of paralysis of the muscular sense.

First we will examine the theory defended by Hitzig and Ferrier, and with some modifications by Carville and Duret.

The hypothesis of motor-centres rests upon two facts which we have already considered and fully verified.

1st. Excitation of strictly-limited cortical zones provokes motion in certain determined groups of muscles.

Respecting the exact limitation of motor-zones, it has been observed by Hitzig, Rouget, and all other authors, that it is only necessary to change the electrodes some millimetres in order to obtain a movement different from the preceding one. According to Bochefontaine,[1] when a section of the convolutions bordering the crucial furrow is examined with the naked eye, conical tufts of the white substance may be seen penetrating the gray cortex, directed towards the surface of the brain. Perhaps as different ones of these tufts are approached, varied effects will ensue upon the anterior or the posterior limbs.

Exact limitation of movement to such or such a muscular group does not prove the existence of a veritable motor-centre for that muscular group. The excitation of any selected sensorial nerve provokes, if the excitation be not too intense, a reflex motion in a certain determined group of muscles. Will it be said that such sensorial nerve is the motor-centre of such muscular group?

To determine the existence of motor-centres then, resort must be had to other proofs.

---

[1] Bull. de la Soc. de Biol., 23d Dec., 1877; Progrès médical, 1878, p. 9.

Pathology cannot supply a rigorous proof of the existence of motor-centres in the cortex cerebri.

According to Charcot, paralysis does not follow a
disease of the gray substance alone, moreover, there are
pathological facts[1] which seem to prove that the localization of motor-centres in various regions of the same convolution, of the ascending frontal for example, cannot yet
be determined with sufficient precision.

Thus it cannot be demonstrated that the gray substance
is a centre.

The experiment which we have before recited showed
that the gray substance was very probably excitable by
electricity. Now there remains to be ascertained what
part electricity plays in the phenomenon of motion.

Couty has made experiments which I have already mentioned ; I cannot enter into their details, but he has noticed
that the gray substance can be anæmied without producing
change in arterial tension · he has concluded that electricity
acts by the white substance and not by the gray. This
conclusion seems to me rather hypothetical, all the more so
that, according to Couty himself,[2] anæmia of the gray substance of the brain produces a constant and considerable
cardiac slowing. It is, then, very probable that the gray
cortical substance acts upon the heart, as seems proven
also by the influence of emotions and sentiments upon the
rhythm of the heart.[3]

As respects the non-excitability of the gray substance
by electricity, we have already dwelt upon the subject
sufficiently to render a return to it unecessary. The very
indirect method of Couty does not seem to have disturbed
the direct and positive proofs of which we have spoken.

Our experiments, as well as those of Franck and Pitres

---

[1] See la thèse de M. Mallebay, des paralysies partielles d'origine corticale.
Paris, 1878, No. 286.

[2] Loc. cit., p. 724.

[3] Voyez la belle leçon de Cl. Bernard sur ce sujet : Leç. sur les tissues vivants,
p. 425.

demonstrate that the gray substance is excitable, but they
do not prove a veritable localization of motor-centres in the
gray substance.

We come to the second fact urged by the partisans of
motor-centres.

2d. Removal of those parts of the gray cortex which are
considered as motor-centres is followed by paralysis.

That paralysis does follow, however, has been contested,
and eminent physiologists, Schiff, Nothnagel, Hermann,
Goltz, etc., have attributed the phenomena observed to
quite another cause.

Without fully entering the discussion, we will notice some
of its particulars.

Schiff[1] says that it is the surety and precision, and not
the energy of movement which is affected. " *Often*," *says
Schiff,* " *we have publicly exhibited two dogs, one deprived of the
cortical centre, the other with both posterior columns of the spine
in the upper dorsal region destroyed, and no difference could be
found in their movements.*"

Ferrier's objection,[2] that loss of the muscular sense does
not exist without an affection of other forms of tactile sensibility, is of little significance. On the contrary, he rightly
says that pathological observations prove that cortical
lesions produce a true paralysis, and not a loss of muscular
sense, and this objection carries great force : we will pres-
ently examine it.

Ferrier also says[3] that that which he calls tactile sensibility, by others termed muscular sensibility, is affected
when the lobule hippocampi is destroyed,[4] but that is
debatable, especially as other observers have detected muscular sensibility from other regions. Hitzig,[5] in particular, thinks that destruction of anterior portion of the sig-

---

[1] Arch. für experim. Pathol. und Pharmac., t. iii., 1874, p. 170.
[2] Loc. cit., p. 350.   [3] Loc. cit., p. 285.
[4] Neue Untersuchungen, 1874, Arch. für Anat., p. 415 et suiv.
[5] Pflüger's Arch. ueber die Verrichtungen des Grosshirns, t. xiii., p. 1, 1876,
t. xiv., p. 412, 1877.

moid gyrus destroys muscular sensibity of the opposite
side, and he concludes:—the convolution upon which muscular sensibility depends is not directly excited by electricity and is not a motor-centre.

We call attention to, as being very interesting, the
recent experiments of Goltz and Gergens to which Hitzig [1]
has as yet but partially replied.   Goltz considers that
which Hitzig calls defect of voluntary energy (*Defect der
Willensenergie*) as only a trouble of muscular sensibility,
and *a propos* gives the following experiment:—With a dog,
habituated to giving equally either paw, the surface of the
left convolutions was removed.   For one month she was
not able to give the right paw ; for two months she made
efforts to do so, but only sometimes succeeded, in a very
irregular manner; finally, at the end of four months she
became able to effect it in a tolerably regular manner,
though with more hesitation than before the operation.
A new operation upon the same side destroyed anew the
power of voluntary movement.   That would seem to prove
that, if the first operation had destroyed voluntary motion,
it was not as a result of a removal of the gray matter or of
the whole substance, but a kind of irritation paralyzing
muscular movements.

Goltz gives another curious experiment.   He made
a continuous pressure, with nippers, to the great toes of
a dog, and observed that the operation, as concerned
paralysis and motor disturbances, was the same that
occurred with a dog in which the opposing hemisphere
had been destroyed.[2] Tripier [3] has seen cases of the same
kind.   With dogs recovered from a cortical paralysis,
a hypodermic injection of morphine, a free bleeding, or
an epileptic crisis brings back the paralysis.   Analogous
facts are observed in man.

The conclusion that this is an exhibition of irritation
(paralysis by arrest, *Hemmungserscheinungen*) is perhaps

---

[1] Arch. für Anat. und Physiol., 1876, p. 692.
[2] Loc. cit., 1877, p. 442.   [3] Revue Mensuelle, 1877, p. 9.

hypothetical. It is, however, worthy of notice, for it again shows the great resemblance that exists between that which is called the motor-zone of the convolutions and the sensorial nerves of the periphery.

It is a similar theory that Brown-Séquard has with well-known talent defended, and for which he has adduced an imposing array of data.

The theory of paralyses by irritation is, however, still hypothetical, especially as the effects of excitation are rarely paralyses, and as by experimentally irritating the surface of the brain, partial epilepsies and contractions can, as before said, be produced.

But the fact that a pathological lesion of the hemisphere produces paralyses and not a loss of muscular sense, should render us very reserved respecting the theory of Schiff, Nothnagel, and Goltz, for here we are forced to recognize a real paralysis with loss of voluntary motion.

As concerns the brain, the comparison between man and animal is not so close and formal as it is respecting many other functions, and of this the best proof is that cortical paralyses in man are permanent, while with dogs they are transitory.

The fact that paralyzed animals afterwards recover has been carefully noticed by Carville and Duret.[1]

These observers also state that, if after a lesion of the sigmoid gyrus (on the right, for example), the animal becoming paralyzed and then recovering, the removal of the left gyrus be effected, it will not result in a paralysis of both sides, consequently it cannot be admitted that a certain part of one hemisphere supplements or assumes the functions of the corresponding part of the other hemisphere.[2]

This brings us to the theory of supplementation, the stumbling-block of the theory of motor-centres.

To judge of this theory, we must include a consideration of two diverse supplemental localities; first, a supple-

---

[1] See Exp. v., p. 434 of the mémoire already cited, also p. 450 et suivante.
[2] See inaugural thesis of Parant : Des Suppléances Cérébrales, Paris, 1877.

ment furnished by the opposite hemisphere, and, second, that furnished by other parts of the same hemisphere.

1st. It is very easy to understand that the right hemisphere might, to a certain degree, supplement the left and *vice versa*, the two being similar and symmetrical. For aphasia, this has been admitted. One may also reason to himself that a dog with the left so-called psycho-motor centre destroyed cannot raise the right foot, but can walk equally well with both feet, the act of walking being a reflex one, provoked by contact with the ground. To explain this fact, which may seem somewhat obscure, let us compare the psycho-motor centre of the retina. Excitation of the retina produces reflex action of the iris. For example, when the right retina is destroyed, the right iris will no longer contract, but in this case, however, if light be suddenly thrown upon the left retina, the right iris will contract by crossed reflex action. These facts are beyond dispute, but experience shows that for cortical centres, supplementation by the opposite side does not exist, and that voluntary motion is retained even after ablation of the gyrus in both hemispheres.

2d. Supplementation by the gray cortex of the same side must be admitted : and this is precisely the supplementation which is most difficult to comprehend.

If the convolution which surrounds the crucial furrow is really the motor-centre of the legs, then, by removing both right and left convolutions, all four legs should become paralyzed ; if not, then it is not a true motor-centre. *A function being given to an organ, a removal of the organ should cause cessation of function.* Both retinas destroyed, sight is abolished. It would then be necessary to admit that there are several organs for one function, several motor-centres for each limb, which certainly is contrary to probability and to fact.

So the term supplementation signifies nothing, except that those who admit it cannot admit true motor-centres. In man, cortical paralyses are not transitory, they are permanent; and, therefore, perhaps a difference should be recognized between the encephalon of man and that of dog.

To sum up :—the ablation of centres, called motor, proves only one thing, which is, that the conduction of will is interrupted ; the same after section of the sciatic, the muscles of the leg are paralyzed ; and still the same after section of the white fasciculi of the cortical motor-zone, the muscles in rapport with these fasciculi are paralyzed.

There is, however, this difference. The conduction of the will is in this last case interrupted only for a time, whereas, when the nerve is cut, it is suppressed forever.

This would seem to indicate that cerebral conduction, at least in the dog, has no absolutely marked-out route. Vulpian has demonstrated that in the spinal cord conduction is carried on equally by all parts of the gray substance. It is possible that the same indifference holds for the brain, though in less degree. *There are habitual roads, but no compulsory ones.*

We will now review the opinion of those who regard the motor actions of the cerebral hemispheres as reflex.

1st. Schiff supposes the reflex power of the gray substance to be annihilated by anæsthetics (chloral, chloroform, ether), without their modifying the conducting power of the white substance. Consequently, whenever anæsthetics suppress a nerve-function, that nerve-function must proceed from the gray substance. Now, anæsthetics suppress the functions of the motor cortex, not only of the gray cortex, but also the subjacent white fasciculi. These subjacent white fasciculi enter the gray centres which the chloroform has paralyzed, and the movements induced by electrization of the cortex are purely reflex.

This opinion includes several hypotheses, which it would be necessary to demonstrate.

FIRST, it is not certain that in anæsthesia, sufficiently profound to suppress all excito-motor power of the convolutions, there is no disturbance in the conduction of the white substance. On the contrary, it is probable that chloroform and ether act energetically upon the cylinder-axes of the white substance, and end by killing them, as they kill the nerve or muscle.

In the next place, can there be no conduction by the gray portions of the medulla-oblongata and bulb, other than such as is identical with reflex action? That conduction by the central axis of the spine is no longer an hypothesis. It has been demonstrated in all ways by Vulpian, and it is very probable that, in the peduncles as in the protuberance, conduction is effected by the gray parts. What is there astonishing, then, in finding the conducting power of these cells abolished by anæsthetics, which kill nerve-cells?

Schiff adds another fact to fully demonstrate that we really have to do with reflex action. It is, that the conduction is very slow, and the retard considerable : very much greater than with the nerves or nerve-fibres, and equally as slow as a reflex action.

2d. In proportion as the excitation is weak or strong, the movements are more or less marked, and at the same time become more and more general.[1]

This is certainly true ; we have at various times observed it. By increasing the exciting current, movements in both legs are produced. Is this a phenomenon of diffusion? Possibly ; but the argument has no great value.

3d. The strongest argument favoring the hypothesis of reflex action is, that along with movements in the limb, there are also movements in the arteries, the iris, the heart, etc. ; consequently, if would be necessary to admit cardio-motor, vaso-motor, secreto-motor centres, etc.

Vulpian, in his course, dwells upon the invariable fact that excitation of the sigmoid gyrus is always sensible. Non-chloralized, the animal always struggles and utters cries of pain : slightly choralized, electric excitation of the gyrus

---

[1] Hermann, Pflüger's Arch., t. x., p. 77.

9

re-wakens them and a profound anæsthesia is requisite in
order that they shall manifest no sign of sensibility. All
the phenomena of respiratory and cardiac arrest are complicated with a painful excitation, and the difficulty is to
discriminate which belongs to sensibility and which to an
excito-motor cause.

It may be regarded as certain that surrounding the sigmoid gyrus there are white fasciculi beneath the gray
substance, the excitation of which produces at the same
time sensation, visceral movements, and arterial contractions. Have all these white fasciculi exact limits, and are
there vaso-motor, secreto-motor centres, etc.? It is very
improbable.

Some light may be afforded towards a theory of these
phenomena from that which transpires in the spinal cord
and its reflexes. The nerves, after they have left the
spine, divide into a multitude of branches, and upon reaching the periphery present a very extended surface. The
peripheric nerve-terminations represent a vast surface
of dissociation, the spinal cord a line of condensation.
Exciting a point of that surface, two series of diverse
movements will result, first, a local movement of the
muscle responding to the touched cutaneous surface;
second, a distant movement which traverses the spinal
cord to the bulb and then is reflected to delicate and im-
pressionable organs, such as the iris, the heart, the vaso-
motors, etc.; these organs, indeed, are extremely perfected
æsthesiometers, and the least bulbo-medullary excitation
provokes them to motion.

This being the case, is not the phenomenon identical
with that resulting from excitation of the brain? There is,
first, excitation of the fasciculi which causes movement in
the muscles; but as that excitation extends also to the
bulb, and as the reflex centres of the iris, heart, and vessels
are extremely sensitive, the iris, heart, and vessels are
immediately influenced by that bulbo-medullary excitation.

Now to a certain degree we can compare the apparatus

of cutaneous peripheric dissociation to the apparatus of peripheric dissociation of the cortex cerebri; the two systems being in close connection with the nerve-axis of the gray substance. The excitation of either one of these apparatuses produces two kinds of reflexes, one limited and localized, depending upon the region primitively excited; the other disseminated, depending simply upon the bulbo-medullary excitation which is irradiated. On the one hand are reflexes of localization, on the other, reflexes of diffusion.

This comparison, however, would not seem to be entirely exact, as pathological anatomy has demonstrated the existence of fasciculi (trophic or excito-motor) which lead directly from the periphery of the convolutions to the spine without entering the central ganglia and the gray substance, the encephalic axis.

It may be conceived that in the white Rolandic fasciculi there are two orders of fibres; some going directly to the spine without entering the gray substance (these are the fasciculi the existence of which seems demonstrated by descending scleroses), others are lost in the central gray substance of the corpora striata, the peduncles, and especially of the protuberance and bulb; their excitation provokes excitation of the bulb, and in this way reflex movements in all the apparatus of organic life. It may be asked if all the white motor fasciculi go directly to the muscles, if a certain number of them do not pass by way of the gray centres of the spinal cord.

All these hypotheses are interesting, because precise researches can probably be instituted to confirm one or another.

If we now bring together these different facts and endeavor to establish a synthesis by applying to them the acquired knowledge of general nerve-physiology, it will be seen that if certain parts of the brain are excitable whilst others are not, it arises from their connections. Faradization produces sensation and pain when the excited fibres are connected with the opto-striated bodies and the bulb, whilst excitation of the other white fibres of the brain not so connected gives no sensation or pain. It also depends upon the connection or non-connection of these fibres with the spinal cord that there are degenerations in certain regions and not in others.

The apparatus of the convolutions can be assimilated to that of the nerve-periphery. They both are joined to the centres by convergent, centripetal fibres, but at their peripheric commencement these fibres are so dissociated that they can be individually submitted to lesion.

It seems that the periphery of the convolutions is the seat of complex actions, actions not understood, *intellection ;* that the excitation of the will is transmitted from it to the ganglionic centres by special fibres, and that from a lesion of these fibres paralysis ensues.

This localization, however, does not exist equally with all animals. Milne Edwards has fully demonstrated that in proportion to the ascent in the scale of beings, division of labor becomes more perfected. Applying this knowledge to innervation, we may suppose that in the same manner conduction becomes more and more precise. The white fasciculi intended to transmit such movements form more numerous and better determined groups. As Vulpian long ago said, and as Ferrier demonstrated in his experiments, the mechanism of the convolutions becomes more important and complex in proportion to the rank which the individual occupies in the animal series. In young animals, which are inferior beings, there is no motor excitability of the cortex cerebri.[1]

The convolutions are an apparatus of luxury, an addition superimposed to the essential vital system, being formed

---

[1] Soltman, Centralb. für d. med. Wissensch., 1875. 14. p. 209. I have been told by Bochefontaine that he has verified this important fact. In new-born guinea-pigs, however, which are at birth quite advanced in development, the cortex cerebri is excitable. Tarchanoff (Gaz. méd., 1878, p. 441), is publishing a very interesting work upon the development of psycho-motor centres in young dogs.

by the substance which surrounds the central encephalo-medullary canal.

As for the facts, they can be recounted in a few lines.

A. In the dog, cortical paralyses are transitory; with the monkey they are permanent (Ferrier), and as concerns man, Charcot has demonstrated by an imposing collection of proofs, that a lesion of the hemisphere without any lesion of the central ganglia will produce permanent paralysis.

B. Cortical lesions (Rolandic) never induce anæsthesia (immediate or consecutive) in man, though they certainly affect sensation in the dog.

C. There is, then, between the human brain and that of the dog a notable difference, so that conclusions cannot be drawn from one for the other, except with great reserve.

D. In man there are some exceptions, very few (if all the doubtful cases are excluded), still a certain number which are well authenticated, where profound and extensive lesions of one or the other of the Rolandic convolutions have not produced paralysis, as also lesions of the convolution of Broca have not resulted in aphasia.

E. In the great majority of cases, however, a cortical lesion produces sclerosis of the lateral cords of the medulla oblongata, etc., without having affected the central ganglia. If, as is probable, the trophic (?) and the excito-motor actions pursue the same tract, it may be concluded that there are white fasciculi going directly from the Rolandic zone to the spine and muscles without passing through the central ganglia.

F. On the other hand, it is certain that, in the dog, any excitation of the Rolandic cortical zone not only acts (directly?) upon the muscles, but also upon sensation, arterial pressure, salivary secretion, the vaso-motors, etc.

G. In the new-born, the cortex cerebri is not excitable, still they are capable of movement and sensation (unconscious?). After abrasion of the sigmoid gyrus, new-born dogs suffer no paralysis.

H. Are there psycho-motor centres? That is doubtful and of very little importance. It is a great result for united physiology and pathology to have demonstrated that the psycho-motor fasciculi are more developed in man than in the dog.

I. 1st. With most individuals, though not with all, psycho-motor fasciculi exist, well limited, and may be individually paralyzed, atrophied, or super-excited.

2d. This individualization of the fasciculi is imperfect in the dog, not only as between the various motor fasciculi, but also between the cortical motor zone and the other zones of the brain.

3d. The psycho-motor apparatus, comprising the cortex of the convolutions and the conducting fibres, is much less developed in animals than in man.

4th. It is an apparatus which is slowly perfected and only in very superior animals.

These conclusions evidently are not very satisfying, but they make no pretense to a theory and only serve to explain how practitioners and physiologists may furnish for the same problem solutions so apparently different.

SEC. 2. SENSORIAL FUNCTIONS OF THE CON-
VOLUTIONS.

We have seen that the motor parts of the cortex cerebri are also the sensorial parts. Vulpian was one of the first to dwell upon this very important fact. This sensibility is that termed, general sensibility.

But in different parts of the cerebral periphery there are still other regions reserved to sensorial functions, especially to vision, and though science is not yet definitely settled concerning the subject, it is Ferrier to whom the honor belongs of having sought these sensorial regions and having to a certain degree determined them.

VISION.—From excitation of the gyrus angularis (*pli courbe*) in monkeys, Ferrier has reported movements in the eyes and head and contraction of the pupils. To ascertain if reflex action was involved, Ferrier abraded the gyrus. Now destruction of the gyrus angularis on one or both sides did not effect motor paralysis, but it produced blindness of the eye opposite to the side of the lesion. If the other gyrus remains intact, supplementation ensues, and after a time a restoration of vision to both eyes is possible. On the contrary, if both gyri be removed, the blindness is complete and permanent.[1]

The modification which this operation effects upon vision it is not easy to state; in reading the details of Ferrier's experiments, however, one is convinced that destruction of the gyrus angularis produces, if not absolute loss, at least a considerable disturbance of vision.

From his experiments Ferrier concludes that when electric excitation of the gyrus angularis provokes movements of the eyes it is the result of a reflex action; the visual centres being intimately united with the motor-centres of the eyes and pupils. As he expresses it, somewhat hypothetically, excitation of the gyrus provokes subjective visual sensations which cause reflex movements in the eyes.

Can Ferrier's opinion, localizing visual sensations in the gyrus angularis, be accepted? It seems difficult, and I find, even in the experiments themselves of that eminent observer, the proof that such localization is not at all possible.

He observes that destruction of the occipital lobes, or the posterior extremities of the hemispheres, disturbs vision : he thinks this effect due to the consecutive inflammation having seriously affected the gyrus angularis, but perhaps he has given that complication undue importance. The fact that monkeys or dogs can be entirely deprived of the posterior part of the encephalon without suffering vis-

---

[1] Ferrier, loc. cit., p. 261 et suiv.

ual disturbance does not prove that those regions have no influence upon vision, provided supplementation is possible, and this fact, whatever difficulties may attend its interpretation, is beyond doubt.

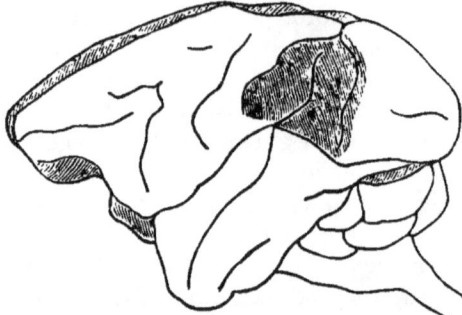

FIG. 19.—Monkey's brain (after Ferrier).
(The shadings indicate locality of lesions which produce blindness.)

In fact, the occipital lobes do have an evident influence upon vision, as Vulpian's experiments at the School of Medicine positively demonstrated.

I will also cite the experiments of Munck,[1] who removed the occipital lobes from dogs and who says that after removing the superior part of the lobes there resulted psychical blindness (*Seelenblindheit*) and from removal of the inferior parts a psychical deafness (*Seelentaubheit*).

I admit that the psychology of Munck, though ingenious, seems to me rather subtle, and that the theory of commemorative images does not seem firmly established. Munck supposes that the removal of certain portions of the dog's occipital cortex destroys the memory of visual impressions, though in some cases certain commemorative images remain in the midst of the loss of all others. " *With one, it was the image of a bucket from which it was accustomed*

---

[1] Zur Physiologie der Grosshirnrinde, Berl. Klin. Woch., No. 35, 1877, p 505. This short notice is translated almost textually in the Revue des soc. méd., 1878, t. xi., p. 33.

to drink ; with another the gesture which asked the dog to give the paw." Aside from these puerilities, however, Munck has stated an interesting fact, that animals blinded by occipital lesion would recover sight and learn again to see. Also, if one eye were removed from a new-born dog, the occipital visual region of the brain upon the opposite side seemed, after a few months, to be atrophied.

Concerning the visual functions of the hemispheres in animals unprovided with convolutions, we have the celebrated experiments of Flourens. These experiments interest us to a certain point : for the cortex cerebri in the higher animals, with convolutions, ought to have analogous functions with the cortex cerebri of lower vertebrata, which have no convolutions.

" *A chicken, deprived of lobes,"* says *Flourens,* " *has really lost sight, hearing, etc. Nevertheless, none of the senses, or rather none of the organs of sense, have been directly injured : the eye is perfectly clear, the iris mobile. Touch, hearing, taste, all the organs of sense are untouched, but the perceptions are lost ; the perceptions, therefore, do not reside in the organs.*" [1]

This psychological physiology seems to me preferable to Munck's.

MacKendrick has also studied the effects of removing posterior and anterior parts of the cortex in the pigeon. Ablation of the anterior part did not injure vision, but ablation of the posterior part caused blindness.[2]

To sum up, it may be considered that visual impressions pass through the posterior occipital and penetrate regions (probably the gyrus angularis) of the cortex cerebri ; but, however probable it may be, it is not proven that conscious visual perception is localized at these points.

HEARING.—Ferrier has endeavored also to ascertain the location of auditive sensation.

He employed two methods, electrization and abrasion.

---

[1] Flourens, Syst. nerveux, 2d Edit , p. 91.
[2] Cited by Ferrier, loc. cit., p. 273.

Electrization of the first temporal convolution in monkeys is followed, Ferrier says, by different results, namely, the opposite ear lowers or rises suddenly, the eyes open widely, the pupils dilate, the eyes and head turn to the opposite direction.

Comparing these phenomena to those observed when the animal is surprised by a harsh noise, Ferrier found them to be identical, and concluded that they arose from the same cause, that is, subjective auditive sensation, this being produced in óne case by excitation of the organs of hearing, in the other by excitation of the sensorial centre.

The movements would be due to excitation of the motor-centres of the ear, which excitation, pursuing a very complex course in the gray cortex, then passing by way of the cortical motor-zone, the corpora striata, the bulb, etc., would finally reach the muscles. Thus the movements of the ear would have reflex origin.

Abrasion of the first temporal convolution seemed to render the animal deaf. The experiment is a difficult one, for great attention is requisite to recognize deafness in animals. Still, Ferrier regards his experiments as decisive. "*Auditive reactions*," he says, "*from electric excitations, and the absence of such reactions from customary forms of auditive excitations when the first temporal convolutions are destroyed, is equivalent to a positive demonstration of the localization of the auditive centre in that region.*"

Munck also locates auditive sensation in the temporal lobes. After the destruction of only one of these lobes he has seen a gradual return of the auditive perceptions. Here then, as in vision, the opposite hemisphere supplements.

TOUCH.—Localizing the sense of touch is attended with more difficulties than surround the other senses, not only because of the uncertainty of knowing whether the animal feels or not, but also for reason that the analysis of tactile sensation is not yet sufficiently complete, and it is almost impossible to distinguish tactile insensibility from reflex and from sensibility of pain.

According to Ferrier, the location of touch is in the gyrus hippocampi. The experiment upon which he bases this localization seems to me far from conclusive, and I will not here recite it. It seems to me that in the monkey upon which he operated there was a certain degree of anæsthesia of the opposite side, that which belongs, as we have just seen, to lesions of the subjacent white substance.

ODOR AND TASTE.—Ferrier's experiments seem to show that the olfactive sensations converge in a definite region of the cortex cerebri, that is, in the cornu Ammonis. Moreover, that localization is in accord with comparative anatomy which shows the connection between the subiculum of the cornu Ammonis and the olfactive lobes.

By exciting the cornu Ammonis in monkeys, Ferrier produced movements in the nose on the opposite side, which might be the indication of subjective olfactive sensation.

With monkeys where the cornu Ammonis was removed, the most disagreeable objects gave no offence to either taste or smell. On closing the opposite side of the nose, it could be seen that the animal no longer had any olfactive sensation.

Broca, by comparative anatomy, reached the same conclusion.[1] Unfortunately I cannot enter into the details given by that eminent master, nor show how close the connection is between the gyrus fornicatus, the olfactive lobe, and the gyrus hippocampi.

HUNGER AND THIRST.—We have seen that according to various authors (Hitzig, Schiff, etc.), muscular sensibility was deranged when the motor regions of the cortex cerebri were injured. This localization is so much the more doubtful as it is not yet certain that muscular sensibility exists.

---

[1] Anat. comp. des circonvolutions cérébrales, Revue d'anthropol., 1878, p.385.

As for the other sensations, called general, they have to
the present been so little studied, on account of the extreme
difficulty of physiological analysis, that their localization
would seem a premature attempt.   I will, therefore, content myself with saying that Ferrier has localized hunger
in the occipital lobes, sexual appetite and thirst in the region of the hippocampus.   But all this is too hypothetical
to detain us.

SENSIBILITY TO PAIN.—The experiments which we have
before recounted, serve in a certain degree to make known
to us the localization of pain.

We have seen, in fact, that excitation of the cortex cerebri in the regions called motor, produced, in animals not
under anæsthesia, reactions indicative of a painful perception, whilst excitation of other regions did not.

We have also seen that upon removing the gray cortex
and exciting the white fasciculi beneath, the same phenomena ensue, and indications also of a perception of pain.

The conclusion to be drawn from these two orders of
facts undoubtedly is, that pain does not reside in the cells
of the peripheric gray cortex, but that in all probability it
is located in the ganglionic axis of the nerve-centres (corpora opto-striata, peduncles, and protuberance).

It should be observed that other experiments led Longet
and Vulpian to consider the pons Varolii as the seat of pain
and the *sensorium commune*.

In reality pain does not exist unless it is perceptible,
conscious, and the seat of pain is probably the seat of consciousness (*sensorium conscium*).

Pathology, which has taught so much concerning motor
functions, has not furnished us yet with a satisfactory solution respecting the sensory functions of the convolutions.
In other words, one cannot produce a chapter of cortical
anæsthesias to place by the side of the existing chapter of
cortical paralyses.

Still, clinical experience and pathological anatomy are
able to indicate cases of hemianæsthesia from cerebral

lesions. These are the cases which, joined to the experimental knowledge that we have briefly outlined, may be sufficient to establish—of course, with such reserve as the obscurity and difficulty of the subject imposes—the part taken by the convolutions in general, or special sensations.

L. Türk, and above all Charcot and his pupils, have fully demonstrated the fact that hemianæsthesia always follows a lesion of the posterior part of the internal capsule, that is, the band of white substance which separates the optic thalamus from the lenticular ganglion of the corpus striatum.[1]

Hemianæsthesia results from a lesion of the internal capsule; consequently the sensorial fasciculi conducting impressions from the periphery to the *sensorium commune* go by way of the internal capsule.

If the *sensorium commune* be located in the cortical layer of the brain, the phenomena observed can be very well explained by a simple interruption of conduction. But if the *sensorium* be located in the protuberance, it must be admitted that the sensorial fibres going to the occipital lobes are reflected, and return by the same way in the protuberance. It cannot be admitted that in man the return is made by way of the motor-zone, for the reason that lesions of the middle and anterior parts of the diverging fibres affect neither vision, hearing, nor sensibility. Perhaps, also, it might be thought that the *sensorium commune* does not exist, and that if the general sensibility resides in the cerebral ganglionic centres, visual sensation resides in the occipital lobes. Again, it might be supposed that in the dog's brain, which is so different from that of man's, the course of sensorial impressions is not the same.

It will be understood why I do not dwell upon these hypotheses any more than I did upon the interesting phenomena of sensorial hemianæsthesia.

---

[1] Charcot : Leçons sur les localisations, 1876, p. 80 et suiv. American ed., p. 63 et suiv.

SEC. 3. INTELLECTUAL FUNCTIONS OF THE CONVOLUTIONS.

I do not believe that at the present time this question can be properly treated. Its solution is probably reserved for the age to come, which, joining the present researches to its own, may be able to establish psycho-physiological laws upon a firm basis.

Still, by availing ourselves of the gifts of embryology, comparative anatomy, pathology, and experimental physiology, we may, to a certain point, establish a relation between intellectual phenomena and the convolutions.

Comparative anatomy shows us that, with mammifera, intellectual activity is in proportion to the development of their convolutions.

Thus, man has very numerous and rich convolutions.

After him, and at a very great distance, come first the anthropoids, then other monkeys ; then the elephant,[1] then the whale, then carnivora.

Rodents, whose intelligence is very mediocre, have no convolutions; they are scarcely indicated.

Dareste supposed that convolutions existed only with animals of large size ;[2] this is but partly true. The castor, for example, has no convolutions, whereas the striped monkey has many ; but what a difference, too, between the intelligence of one and the instinct of the other!

Thus, from their connection, we may suppose that intelligence is a function of the convolutions. It should be added, that from the rodent to man, not only the convolutions augment in number and extent, but also the brain augments in weight. If the proportion between the weight of the brain and that of the body be noted, it will be seen

---

[1] See plate of Leuret.   [2] Ann. de sc. nat., 1865, t. iii., p. 65.

that in man the body weighs 45 or 47 times more than the brain ; with the horse it is 400 times greater, with the elephant 500, the ox 800 (Leuret).

Study of the convolutions, according to age and race, leads to analogous conclusions. The weight of the brain and the richness of the convolutions increase with intelligence.

Concerning the weight of the brain, here are some figures by Davis:[1]

21 English,		.	.	1,425 gr.
25 Chinese,		.	.	1,357 "
5 Esquimaux,		.		1,396 "
9 Negroes,		.	.	1,322 "
17 Australians,	.	.	.	1,197 "

Very accordant results have been observed by other authors, and it may be regarded as certain that, in the white race, the weight of the brain, either absolute or relatively to the body, is greater than in other races.

There is also a difference in the forms of the convolutions.

For comparable elements, it will be necessary to take the convolutions of anthropoids, inferior human races, fœtuses, idiots, and superior races.

In monkeys, even of the highest order, the gorilla for example, the convolutions are very undeveloped as compared with man's.[2] Still, the two brains are constructed absolutely upon the same model, and of all animals the anthropoids have the best developed convolutions.

The same may be said respecting the brain of the Charruas, which is nearly simian. By the following cut, borrowed from Leuret, it can be seen how simple and elementary the convolutions of the parieto-temporal

[1] Cited by Pozzi, Revue critique sur le poids du cerveau ; Revue d'anthropol., 1878, p. 277. I refer to that article for bibliographic and other details into which I cannot enter.

[2] See pl. ii., fig. 2 in Broca's mémoire : Sur le Cerveau du Gorille. Revue d'anthrop., 1878, p. 1.

regions are, especially at the point where the parallel fissure becomes buried in the parietal lobe. The temporal convolution which ascends, to form the convolution of the gyrus angularis, is extremely simple, without annectant or transition gyri, and if the brain be compared with that of a European, the difference between the simplicity of the one and the complexity of the other will be striking.[1]

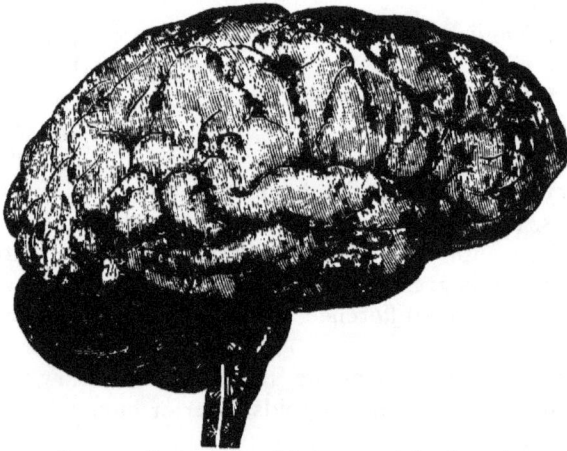

FIG. 20.—Brain of the adult Charruas (after Leuret).

It should be observed though, that in some Europeans, feeble in intelligence or imbecile, and in some criminals, there is the same exterior formation, the same simplicity of convolutions. In the accompanying figure (fig. 21) of the brain of Fieschi, also borrowed from Leuret, can be seen the extreme simplicity of the temporal lobe.

The sulci which separate the three temporal convolutions are straight, parallel, without flexures; to a certain degree it is of the simian type.

To these may be compared Pozzi's excellent illustration of the brain of the imbecile, Marie Martel.[2] The poste-

---

[1] For example, see fig. 440. Traité d'anat. Sappey, p. 63.
[2] Art. Convols. du Dict. Ency., figs. 6, 7, and 8, pp. 352, 354, and 355.

rior parietal and temporal convolutions are extremely
simple. It may be seen also that the frontal convolutions
are straight, simple, and narrow, and the paucity of folds
is general, equally marked in all the lobes and all the convolutions.

It is interesting to compare with these primitive brains
the brains of the fœtus or of new-born infants, in whom no
intelligence yet exists. In these, the secondary convolutions
are not developed. It is only at about the age of puberty
that the organ is entirely developed and its functions established.[1]

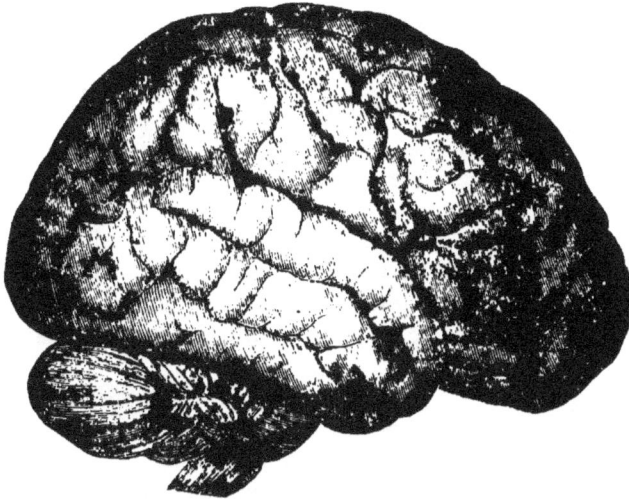

Fig. 21.—The brain of Fieschi (Leuret).

If we add to this the astonishing resemblance which
some microcephalic brains bear to the brains of monkeys,[2]
it may perhaps be concluded that intelligence and richness

---

[1] See fig. 21 of Ecker, in art. of Pozzi, loc. cit., p. 384.—Also Luys, Bull.
de la Soc. de Biol., 1876, p. 230, Cerveau d'une imbecile: Lebon, Comptes
rendus de l'Acad. des Sciences, 1878, 8 juillet.
[2] See pl. xxiv., Leuret, Atlas, p. 36, figs. 4, 5, and 6.

of convolutions, more particularly it may be of the posterior parietal and occipital, maintain a close relationship.
It is possible also that the frontal convolutions are more
developed in proportion to the degree of intelligence. It
seems, indeed, that intellectual labor appears to increase
the volume of the frontal lobes.[1]

An attempt has been made to establish a difference between human races based upon developed foreheads as
representing intelligence and developed occiput as a mark
of inferiority, but the distinction is not established, as the
development of the occipital regions which encroach upon
and cover the cerebellum is one of the features which
most distinguishes the human brain.

As Broca said, symmetry of the convolutions seems to
indicate an inferiority.[2] The human brain is much more
asymmetrical than that of the monkey or of animals. In
idiots, microcephales, etc., the hemipheres often are symmetrical. At the same time, the brains of idiots often are
very asymmetrical.

As a rule, the female brain is less rich in convolutions
than the male, and its weight, as compared to the weight
of the body, is also less: here, too, intelligence and the
greater or less development of the convolutions are in
unison.

In short, the brains of those feeble in intellect and inferior in type have less weight and greater poverty of
convolutions. It is the reverse in brains of those highly
intelligent. Cuvier's brain weighed 1,829 grammes; those
of Cromwell and Byron weighed still more, if figures not
very well authenticated can be accepted.[3]

Recent autopsies, the only ones possessing any value, as
it is only since Leuret and Gratiolet that the morphology

---

[1] Lacassagne et Cliquet, De l'influence du travail intellectuel sur le volume
et la forme de la tête.—Bull. de la soc. de méd., pub. 1878, p. 398.

[2] Bull. de la Soc. d'anthrop., 1866, p. 393.

[3] See Wagner, Recherches sur les fonctions du cerveau, Journ. de la physiol.,
iv., p. 554.

of the convolutions has been known, leave us almost completely in default, and the question may be considered as
still under consideration.[1]

It is possible that the relation between intelligence and
the measure of convolutions and weight of brain would
not present the seeming irregularities observed, if the convolutions of the brain could be unfolded, as Gall proposed.
I am not sure whether that unfolding could be effected
even to an approximative degree ; but that would be an
important element of knowledge, in order to judge of the
quantitative value of the gray substance of the brain.

Again, perhaps it is not only the quantity, but also the
quality (?) of the gray substance which plays an important
part.

Let us see now whether experimental physiology and
pathological anatomy will furnish more positive ideas respecting the intellectual functions of the convolutions than
comparative anatomy does.

Experimental physiology reveals little ; the most precise
knowledge thus far from this source is derived from the
admirable experiments of Flourens. When the cerebral
lobes are entirely removed from birds or reptiles, vision,
hearing, and intelligence are abolished. They no longer
possess volition ; their movements are reflex and involuntary ; and although they are appropriate to the result, still

---

[1] " The weight of Sir Jas. Simpson's brain (1870), including the cerebellum,
was 1,530 grammes. The convolutions were remarkably numerous, besides
which, they were bent and interlaced as though the skull afforded insufficient
room. The island of Reil was astonishingly developed." Revue d'Anthrop.,
1872, p. 124. Dante's skull was remarkable for the enormous size of the
frontal lobes (Broca, Bull. de la Soc. d'anthrop., 1866, p. 206).—Gratiolet also
mentions the marvellous form of Descartes' head (ibid., 1861, p. 71).—For
descriptions of brains of idiots or imbeciles, I will specially refer to the Journal
of Mental Sciences and to the West Riding Lunatic Asylum Reports.—Luys
has called attention to a supplementary convolution, which, lying parallel to
the post-Rolandic, joins the temporal to the superior parietal lobe ; but it is
not certain that the existence of that convolution coincides with greater intellectual activity. Bull. de la Soc. de Biol., 1876, p. 222.—See also Vulpian's
lessons upon the Physiologie du système nerveux, p. 688.

they in reality are entirely regulated by the intensity and nature of the excitation.

A great number of all kinds of experiments have been made upon the convolutions of mammifera, but the conclusions are still uncertain and obscure.

The Rolandic zone can be destroyed without altering the intelligence; also, a considerable portion of the frontal, or even the occipital lobes, can be removed without an apparent alteration of the intellectual functions; but a decision is so difficult, and the interpretation so debatable, that as yet there is nothing established in this mysterious question. It is possible, however, that the blindness and deafness, caused by destruction of the occipital lobes, and perhaps, also, the ruining of other ill-comprehended functions, may throw the animal into a state of stupor nearly equivalent to the loss of intelligence.

If all the considerations hitherto presented in this work be taken into account, we will be forced to recognize that animal physiology is incapable of resolving the question ; for concerning intellectual faculties, no comparison whatever can be instituted between man and animal.

To elucidate the two following problems, then, pathology should be interrogated :

1st. Is intelligence located in the convolutions?

2d. In what convolutions are the various intellectual faculties located ?

For man, life without the brain is impossible, though there are numerous cases of the destruction of one lobe (frontal, occipital, or parietal), or of two corresponding lobes (the two frontal for example), without destruction of intelligence.[1] It may be said that instances are not infrequent where the frontal lobes have been traversed by a ball, and yet the intelligence retained, or rather, nearly so.

Experiments upon animals give the same results, and the

---

[1] Among other cases, one cited by Vulpian, loc. cit., p. 70. See also Mire, Rec. de méd. militaire, juin, 1871.

following proposition may be deduced; one that thus far has not been invalidated.

There is no region upon the surface of the convolutions especially assigned to intellection.

Still, there is evidently one localization of intellection: aphasia has served to localize the function language, and perhaps that will make possible an explanation of the nature of intelligence.

Intelligence seems to be the union of various faculties, and language is one of those faculties. Thus the intelligence of an aphasiac is impaired in one of its constituent elements, and there is an intellectual diminution. A man with his mouth closed might have all his memory, judgment, and imagination. That man would not be an aphasiac; on the other hand, observation shows that there does not exist an aphasiac with full memory, judgment, and imagination. Perhaps there may be other faculties localized, like language, in a convolution. But what are the faculties which, gathered into a fasciculus, constitute human intelligence? Can we hope to find another faculty so well defined psychologically as is that of language?

We will enter no further upon this subject which is too metaphysical and too open to debate; we will only add a few words respecting general paralysis of the insane.

Since Calmeil and Parchappe, it is known that the anatomical lesion in this affection is an alteration in the gray cortical substance, according to modern authors, in the deep [1] layer of it.

Would that region be really the seat of intelligence? It is very hypothetical. All that can be said is, that very probably the gray cortex of the brain is the organ of the intellectual functions.

With general paralytics of long standing, having arrived at the last degree of dementia, and in senile dementia, the gray substance is found to be atrophied, invaded by conjunctive tissue, having in fact almost disappeared.

---

[1] Meschéde, Allg. Zeitsch. für Psychiatrie, 1873.

It is probable that accesses of delirium in general paralytics coincide with extensions of congestion. It is known
to the medical attendants of the insane that the delirium
of congestion is an exalted kind [1] (*ambitieux*).

There is a manifest relation between the quantity of
blood which circulates in the superficial layers of the brain
and the conception of certain delirious ideas.[2]

Again it is known that alterations of the intelligence
from poisons, that is to say, functional disturbance in the
gray cortical substance, first affect the voluntary conscious
faculties, and afterwards the involuntary, unconscious ones.
It is not amiss to recall the fact that the same substances
which abolish reflex action (chloroform, morphine, alcohol,
etc.), are also the substances which abolish intelligence.
Poisons to the gray substances of the spine (poisons of the
reflexes) are also poisons of the gray cortical substance
(poisons of intelligence). The fact that a few drops of
alcohol introduced into the circulation produce furious
delirium explains the reason why, in many cases of short-
standing insanity, no cerebral lesion can be discovered,
either by the naked eye or with the microscope; it is
from the reason that, besides an alteration of form, there is
probably also a dynamic, functional alteration which is
beyond our present means of detection.

If in a few lines we would now embrace a conclusion of
this work, too hasty to escape imperfection : we would say
that the brain is not a simple organ, and that it would be a
false path to seek therein a general focus uniting all im-
pulses, impressions, and volitions.

The inferior vertebrata are very simple beings; their
movements are nearly automatic and seem to be reflexes

---

[1] Perhaps there should be some reserve respecting the congestion of paralytics. See Charcot, Leç. sur les maladies du système nerveux, t. i., p. 251.

[2] Moreau has often told me of a cobbler, a general paralytic, who, during congestion, believed himself the pope. Bleeding calmed the congestion and stopped the delirium, but whenever the flow of blood was stopped, by placing the
fingers over the opening, he again became pope.

of the least possible complication. But as one ascends the scale, a perfecting apparatus becomes added to that primitive, almost embryonic system. That apparatus is the cerebral gray cortex. The more the psychical, sensorial, and ideo-motor functions are developed, the more the apparatus which is destined to their use also acquires development. In the superior mammifera this layer of nerve-substance has to assume folds and irregular volutions in order to find room in the cranial cavity. It is in this layer that the intellectual functions are elaborated, and from thence also come the psycho-motor impulsions.

The route taken by these impulsions is now known ; it is by way of the white fasciculi neighboring the fissure of Rolando.

Along with these motor impulsions there are others, probably involuntary and unconscious, located in the ganglionic axis surrounding the central canal.

It can be imagined that in the superior mammifera, especially in the monkey and in man, there is a kind of dual existence ; the one a simple reflex with which the cortical apparatus has nothing to do ; the other a complex reflex, intellectual, in which the sensorial, conscious impressions, passing through the occipital layers, become elaborated by the entire cortex cerebri and then returned by way of the parietal white fasciculi as voluntary motor-excitations.

But even this is as yet hypothetical and long researches are still requisite to build up (or destroy) the theory.

Everywhere are found so many hypotheses and so few certainties that discouragement might well ensue upon beholding the colossal efforts of an age of research producing results so contestable. But such discouragement would be unjustifiable, for each day brings progress, and the moment will come, perhaps is near at hand, when the functions of the convolutions, that is to say, intelligence, will be understood, as are the functions of the heart, the muscles, and the blood.

END.

BIBLIOGRAPHY.

A number of critical reviews, physiological and pathological, have appeared upon the question of localizations, and also some interesting works not directly related to physiology.
I subjoin a list of some of the principal ones :

DODDS. Localization of Functions of the Brain, being an historical and critical analysis of the question. Jour. of Anat. and Phys., t. xii., p. 340.

RENDU AND GOMBAULT. Des localisations cérébrales. Revue des sc. méd., t. vii., 1876, pp. 236 and 765.

LEPINE. Revue mens. de méd. et chir., mai, 1877, p. 381.

BOURDON. Recherches sur les centres moteurs des membres Bull. de l'Acad. de méd., 23d Oct., 1877.

POZZI. Des localisations cérébrales et des rapports du crâne avec le cerveau au point de vue de la trépanation (Arch. gén. de méd. Ar., 1877).

GLIKY. Ueber die Wege auf denen die durch elektrische Reizung der Grosshirnrinde erregten motorischen Thätigkeiten durch das Gehirn hindurch fortgeleitet werden (Eckhard's Beiträge, etc., t. vii., p. 177.)

VETTER. Ein Ueberblick ueber die neueren Experimente am Grosshirn. Deutsch. Arch. für klin. Med., xv., p. 350.

CZARNOWSKI. Ein Beitrag zur Lehre von den motorischen Centren der Grosshirnrinde. Diss. inaug., 32 p., Breslau, 1874.

ONIMUS. Des erreurs qui on pu étre commises dans les expériences physiologiques per l'emploi de l'électricité (Gaz. hebd., 1877).

SEGUIN. A Contribution to the Study of localized cerebral Lesions. Rep. from the Trans. of the Amer. Neurol. Assoc., New York, 1877.

BORDIER. Revue critique des localisation cérébrales.—Rev. d'anthrop., 1877, p. 265.

GRASSET. Des localisations dans les maladies cérébrales, 2d Edit., Montpellier, 1878 (a very good and complete work).

ATKINS. Revue sur les localisations. Dublin Jour. of Med. Science, July, 1877, p. 50.

NEWCOMBLE. Epileptiform Seizures in general Paralysis. West Riding Lun. Asylum Reports, 1875.

A. FOVILLE, son. Des relations entre les troubles de la motilité dans la paralysie générale et les lésions de la couche corticale des convolutions fronto-parietales. Ann. méd. psychol., Dec., 1876, Jan., 1877.

MATHIAS-DUVAL. Localisations cérébrales dans les hémisphères. Trib. méd., 1877, p. 248.

BURCKHARDT. Des centres fonctionnels du cerveau. Zeitsch. für Psych., 1877.

FERRIER. Lectures on the Localization of Cerebral Diseases. Brit. Med. Jour., 1878, Nos. 899, 904, p. 399, etc.

TAMBURINI. De l'état actuel de la physiologie normale et pathologique de l'intelligence. Lo Sperimentale, févr. 1877.

OBERSTEINER et EXNER. Mesure de la vitesse de la pensée chez les aliénés. Arch. für Pathol. Anat., 1873–1874.

BESSER. Reflexe der Neugebornen. Arch. für Psychiatrie, t. vii., p. 460.

BALOGH, HILAREWSKI, HOLMGREN, WELIKY. Being various works in Hungarian, Swedish, and Russian, analyzed in the Jahresbericht für Anat. Physiol., pp. 35 to 40, 1876. Hilarewski holds that the convolutions have no vaso-motor action; on the contrary, Balogh considers that the hemispheres influence the cardiac rhythm. Lastly, Weliky limits diffusion to three millimetres.

POISONS OF THE INTELLIGENCE.

POISONS OF THE INTELLIGENCE.

ALCOHOL, CHLOROFORM, HASCHISCH, AND COFFEE.

I.

M. CLAUDE BERNARD defines a poison as a substance which cannot enter into the composition of the blood nor penetrate into the organism without causing transient or permanent disorders. Thus we distinguish a poison from an aliment, since the latter is an assimilable substance that should form part of the blood or other tissues, while the former should be eliminated and disappear. All medicines are poisons and the converse, if we use the word poison in its proper and scientific sense. The early experiments with poisons were merely for the purpose of clearing up medico-legal problems; now we recognize experimental toxicology as part of the study of therapeutics. A great advance was made in the attempt to limit the action of poisons to single organs or tissues, and this progressive period may be said to date from the brilliant experiments of M. Claude Bernard upon the action of curare.

The principle of life resides in no particular organ or tissue, but is disseminated. A living being is one that is composed of living organs which can die separately. These organs are composed of tissues, and these again of cellules, and all may disappear successively without the death of one being necessarily followed by that of the others.

Carbonic oxide acts on the red globules of the blood, and death from charcoal-poisoning is the consequence of the poisoning of this particular anatomical element by carbonic oxide. Therefore the blood dies before the other tissues,

and if these finally perish it is because they are deprived
of living blood which is essential for their existence. The
same process occurs in death from hemorrhage. Physiological analysis goes even farther than this, for not only
are the red globules acted upon, but a particular constituent of the red globules, hæmoglobine.

Unfortunately the same accurate knowledge does not
extend to the action of all poisons. We know that the
blood, muscles, spinal cord, and nerves have properties
that are perverted or destroyed by special poisons, but our
information rarely goes beyond this primary localization.

By poisons of the intelligence we mean those whose primary action affects the processes of intellection. This
may not be their exclusive action, for other organs and
functions may suffer afterwards, but it is the predominant
action. The intelligence is attacked before any other functional troubles appear, but these must follow, for the central nervous system presides over muscular movements
and the functions of all the organs and apparatus of the
body, while at the same time it is the organ of intelligence.
So chloroform, which begins by suppressing the will, memory and ideation, functions of the brain, ends by paralyzing the movements of the heart and respiration, functions
of the spinal cord.   Strychnine acts in a reverse way.

If we consider the nervous system as having three principal functions, intelligence, which depends upon the
cerebrum, voluntary movements, depending upon the spinal cord, and organic movements of the heart, digestive
apparatus and glands, depending on the medulla oblongata,
we shall find some poisons acting first on one or the other
of these parts and upon the functions which they control,
though finally upon all parts of the nervous system.

We shall consider only those which act upon the brain
and disturb the intellectual functions.   We shall not seek
to determine how they act, since this is unknown, but endeavor to give as clear an analysis of their symptoms as
possible.

II.
ALCOHOL.

The chief characteristic of all poisons of the nervous system is that they excite before they destroy, and it is the pleasure of this sur-excitation that is habitually sought by drunkards. The first effect of alcoholic intoxication is a secret sense of satisfaction, a most agreeable beatitude. The ideas seem to clear up, difficulties vanish, life is rose-colored, and full of contentment and happiness. If one continues to drink, the intellectual excitation augments and shows itself in various ways. We may sum up all these forms in one word by calling it a condition of hyper-ideation.

In this condition there is a profusion of ideas and fancies of the most varied and opposite character which succeed one another with great rapidity. That which distinguishes them is their lack of moderation. There is no measure for the intelligence, all is, out of order and expansive. They feel the moral forces increased tenfold, they think themselves capable of undertaking anything and accomplishing anything. Meantime new ideas follow each other without cessation; all are impracticable, but all amuse for the moment until they vanish and others appear. Perhaps in the number there is something rational, but they have no time to grasp it. There is a perpetual to-and-fro motion of the fancy, a mental ' will-o'-the-wisp ' dance, more or less seductive, in which they cannot find time to make a pause. In this condition it is impossible for them to keep secrets, they become confidential and effusive. This tendency is manifested even in a slight degree of intoxication, but in a more advanced condition there is no confidence that can be kept. This hyper-ideation is chiefly an excess of imagination, and there are some authors who cannot

write unless in this condition of sur-excitation which gives
to their works a factitious stamp of originality.

Often while intoxicated, in the midst of the deluge of
ideas, all of a sudden there appears, without any logical
association, one idea which has nothing in common with
the preceding fancies, and which fixes itself with desperate
tenacity and constantly recurs amidst the others, just as
in concerted music the theme constantly appears among
the modulations and variations surrounding it.

So we find two special characteristics of the first stage
of intoxication : one is a rapid succession and the other a
certain fixity of ideas. There is no contradiction between
these two forms, if we examine carefully the mechanism
of the intelligence.

In the normal state all the faculties, imagination, memory,
judgment, the association of ideas, are governed by
another superior faculty, the attention. Attention, or the
will, is the man himself. It is the individuality which,
being in full possession of resources which it controls,
uses them when and where it wills. Now, even in the
beginning of intoxication, the will and attention disappear.
There is little more than imagination and memory left,
which, abandoned to themselves without guidance and
control, produce the most unexpected results. Sometimes
there is one idea that cannot be driven away, sometimes another that cannot be retained, for attention is
as effective for the elimination of certain ideas as for the
fixation of others. The fixed idea, then, may depend on
the absence of attention fully as much as the too fleeting
fancy, and in both cases it is the direct result of poisoning
of the brain by alcoholized blood. Therefore, although
in the first stage of intoxication it may seem that one's
ability for work is increased, he will soon find, if he really
wishes to work, that he is unable to collect and fix his
ideas, and the delusive fertility with which he thinks
himself endowed will very soon appear to him like an
actual impotence against which he cannot struggle.

Sometimes, however, by chance or from habit, the idea that is involuntarily fixed, is precisely the one he wishes to elaborate, and this fortunate coincidence may convince him that his attention is intact. But this will prove to be an illusion, for it will be impossible for him to do any other work. It is this loss of attention, then, and the sur-excitation of the imagination as well as the diminution of judgment, that characterize the first effects of intoxication.

There are some men, however, who cannot be made tipsy. After a great quantity of alcohol has been taken they will finally have all the symptoms of profound intoxication, but they will not have had, in appearance at least, that period of intellectual excitation that characterizes the first moments of intoxication. This peculiar phenomenon is due to the influence of the will. In these cases will and atten-tion, although diminished, have not entirely disappeared, and the will is even concentrated on the fear of intoxi-cation. Thanks to this fixed idea, which the intoxication exaggerates still more in intensity, there is no external manifestation of delirium. Although the psychological effects of alcohol may be felt, there is yet power enough in the will to restrain its exhibition to others. When, on the contrary, they abandon the will-power by giving loose rein to all the ridiculous fancies that occur to the mind, they are then no longer able to leave off, and it will require a grave emergency to put a stop to the hyper-ideation and the overflowing speech. If one refuses to act upon the first development of ideas, by so much he remains master of himself so far as his tongue and his acts are concerned. It is characteristic of those who consent to become tipsy, who say to themselves at the beginning of a dinner, 'we will be free and easy,' that with the very first glasses they are drunk. Sometimes the intention is equivalent to the act itself, and one may become drunk without drinking. This effect may be produced by good news, unexpected fortune, unhoped-for success, producing results analogous to intoxication, so that in common parlance we say such a

11

one is 'drunk with success.' This moral intoxication, which closely resembles the sur-excitation of alcohol, is, however, rarely observed. There are certain persons whose nervous temperament is delicate and excitable. They are nervous so far as the brain is concerned. The slightest accident upsets their judgment, the smallest emotion or least annoyance destroys at once their presence of mind and courage. In their normal state they lack neither judgment nor will-power; but let an unforeseen accident happen and they lose their heads, and their condition is equivalent to that produced by intoxication. In such persons the slightest access of fever brings delirium. They are weak-headed, and unless they are careful they may become intoxicated with deplorable facility. This predisposition resembles hysteria.

Besides the predisposition of the individual, there are other things which modify the effects of alcohol. There is a difference relatively between the rapidity with which various liquors affect the intelligence.

The intoxication of brandy is dull and heavy. It produces scarcely any intellectual excitation, and seems at first to act upon the organic functions of circulation and respiration. The intoxication of wine, however, is light and stimulating, particularly Champagne and Burgundy, which are noted for their psychic effects. The mixture of liquors seems to increase the intensity of their action to a great degree.

The rapidity of alcoholic absorption is a matter of importance. When one is fasting, alcohol acts very quickly; after a hearty meal it is absorbed more slowly. A queer custom is said to have prevailed among the heavy drinkers of England in former times, and this consisted in drinking a glass of oil at the beginning of dinner and thus preventing the absorption of alcohol by the stomach and intestines. Physiological, but disgusting. Intense heat seems to be effective in rendering liquors more intoxicating than usual. In Egypt, I found a slight quantity of Bordeaux

wine and water too stimulating to be used without great
care. The sudden effect of cold, by checking the excre-
tion of alcohol through the lungs and the perspiration,
may produce immediate intoxication in those who have
taken alcoholic drinks. Hence the monks of St. Bernard,
according to Dr. Burrill, give only coffee to travellers.

It appears then that a small dose of alcohol sur-excites
certain intellectual faculties, while it paralyzes others, but
if the dose be repeated or if the amount is sufficient in the
first instance, there may be a total disappearance of all
trace of the intelligence. The person is dead-drunk. There
is complete anæsthesia, a true coma. Now between these
two periods of sur-excitation and coma occasionally there
occurs a very serious condition which the ancient authors
called "convulsive intoxication." The man is mad, a perfect maniac in appearance and action, dangerous to himself and to others. No frenzy can be more ferocious or
appalling. It is to this period that the crimes and murders
belong that are committed by drunkards. It occurs only
in those whose blood is vitiated by previous alcoholic ex-
cesses, and this furious delirium may come after a new
excess in drink that is relatively less than those that have
preceded it.

There is another form, the chronic intoxication, which
profoundly disturbs the functions of the organs, and ends
by altering their tissues. The nervous system, and more
particularly the brain, is altered more perhaps than other
organs. In the case of dogs whose food has been mixed
with alcohol, we find the brain has absorbed a definite
amount of alcohol, which may be recovered by distillation.
If they are fed in this way for a long time, they become
restless and melancholy, and end by losing their senses.
According to M. Magnan, they have hallucinations; thinking themselves pursued, they run off affrighted, howling
and snapping at the empty air.

In man, likewise, melancholy and fear are the results of
chronic poisoning by alcohol. By a legitimate penalty,

Nature makes expiation for the joys of intoxication by the terrors of alcoholism. At first there is only a vague sense of melancholy which they seek to combat with new doses of the poison. Little by little this melancholy increases. At night, while half asleep, phantoms, ill-defined but repulsive, may appear. These are merely illusions, not yet so vivid as to be called hallucinations. But soon the hallucinations come, hideous forms and frightful creations of a diseased brain. This form of delirium is amply illustrated in medical literature. Sometimes these hallucinations are so frightfully appalling that they compel the unfortunate victim to suicide. According to Brierre de Boisemont, in a total of 4.595 cases of suicide, 530, or about one-ninth, were attributed to alcohol. Poverty and drunkenness are in all countries associated, and we must attribute the use of alcohol by those who are constantly struggling against hunger and cold as due to an instinctive desire for forgetfulness of past and present evils, a wish to blunt the sensibilities. It is, therefore, among the poor, and especially in cold climates where poverty is harder to endure, that alcoholism makes its greatest ravages. In England, among a million paupers helped by public charities in 1865, there were 800,000 drunkards. Whiskey and gin were the most common drinks, together with porter, ale, and stout. In the United States, where the thirst for alcohol is hardly less than in England, besides whiskey and gin, adulterated brandy and rum are added to the list. In Sweden, where alcoholism has made great ravages, according to statistics, each inhabitant, excluding women and children, consumes about one hundred litres (22.01 gallons) of alcohol annually. In Russia the consumption is enormous. They use, besides the spirits of grain, such as vodke and kummel, brega or white beer, symorosli or birch-wine, and a great number of other alcoholic drinks. The Tartars of the East drink fermented mares' milk, a very alcoholic liquor known as koumys.

In temperate climates and in the south of Europe drunkenness is a rare vice. Finally, excess of drink destroys every year about 50,000 persons in England, 40,000 in Germany, 25,000 in Russia, 4,000 in Belgium, and 2,000 in France. People reduced to servitude, or who have emigrated in order to support themselves, are rarely sober. The Irish and Poles are examples of this. The Chinese in their own country are a very sober people, but when they go abroad they become desperate drunkards.

Savages brought in contact with advancing civilization invariably borrow its most vicious features, and acquire at once the habit of intoxication.

Among the alcoholic poisons we have not yet mentioned absinthe. In fact, absinthe does not act solely by means of the alcohol it contains, but rather by the essence of absinthe, which even in a small dose is a notable poison. There is this difference between absinthe and alcohol, that instead of acting solely on the encephalic nervous system, absinthe also acts with great rapidity upon the spinal cord, producing tremors, paroxysms of epileptiform convulsions, and finally attacks of true epilepsy. The surexcitation of absinthe is also more marked than that of alcohol. In every case it is an energetic poison, and its prolonged use is far more injurious to the intelligence than that of alcohol, as has been shown by the researches of M. Magnan.

Alcohol is an excellent stimulant, and in moderate doses is useful as well as agreeable. Its effect upon the nutritive process is well known, and its tonic action is incontestable. But how feeble are its advantages contrasted with its evils!

III.

CHLOROFORM.

Chloroform should be classed with alcohol. Physiologically the action of these two poisons is nearly the same, and although their employment is different, their function is nearly identical.

The principal action of chloroform is the paralysis of sensation, or anæsthesia. This means that it acts upon the intelligence, for sensation is only one of the forms of intelligence. This point, however, needs some explanation.

Two grand functions are vested in the nervous system, sensation and movement. It is by sensation that we receive impressions that come from without ; it is by the excitation of the muscles, or movement, that we manifest our will and act upon exterior objects. When one is neither sick nor poisoned, the will (that is to say, the intelligence) excites, through the medium of the spinal cord, the different muscles, and produces movement. But this condition is not absolutely necessary, since, in decapitated animals, for instance, the nervous system of the spinal cord may still excite muscular movements. There is motility, but no sensation. There is sensation only when the intelligence is intact and capable of perception ; therefore, when there is no intelligence there can be no sensation. This is supported by pathological observations. Whenever the intelligence is attacked, there are at the same time disorders of sensation, and reciprocally. So when we see a patient presenting notable troubles of sensation, the nerves being intact, there must be a lesion of the central nervous system, and a lesion which does not leave the intelligence unaffected.

Anatomy and comparative physiology are in accord with pathology. Some animals have very little perception ; they are the inferior animals, their intelligence is obscure, and their sensation is as obtuse as their intelligence. But in proportion as we consider the more intelligent animals, we see sensation becoming more and more delicate, until in man, the most intelligent of all, sensation is the most perfect. So among the different races those are the most intelligent who are the most sensitive. The anatomical arrangement of the nervous centres is in accordance with this coincidence. In man the posterior columns of the spinal cord are more voluminous than the anterior columns. Now the anterior columns transmit the motor excitations to the nerves, while the posterior columns serve for the conduction of sensitive excitations. In like manner the posterior lobes of the brain, compared with those of animals, are relatively more developed in man than the anterior lobes. It is in the posterior lobes that the perception of sensitive excitations, or the faculty of sensation, appears to reside.

Whatever may be the spontaneous development of the mind itself through the proper constitution of its organ the brain, all our knowledge comes from our sensations and from the brain-work depending on them. Intelligence is in some sort the product of these two factors.

Poisons which act upon the intelligence are, therefore, poisons of sensation, and in this respect alcohol does not differ from chloroform. In the beginning of intoxication there is already a marked insensibility, but in the comatose period insensibility is absolute, just as in the last period of chloroform poisoning; so that chloroform intoxication follows a march parallel to that of alcohol, and we may distinguish a first period of chloroform intoxication and a second period of sleep or coma.

When one inhales chloroform, the first whiffs make him giddy and create a very disagreeable sensation of vertigo and bewilderment. This vertigo augments, and in proportion as the patient continues to inhale the toxic substance his ideas become exalted. He understands and responds to all that is said, but his replies are like one intoxicated; his impressions are exaggerated, his judgment is gone, and he gives a theatrical accent to the most insignificant responses, producing a grotesque effect. The ideas become more and more confused, will and judgment are abolished, imagination becomes disordered and delirious, in a word, it is a condition of sleep with dreams, nearly analogous to that which occurs in ordinary sleep.

Attention, judgment, the will, and memory disappear at the same time, so that we may see the peculiar spectacle of a person living and thinking, but whose life and thought leave no trace on his memory. When we tell him what he said or did, it will be news to him. There will have been a lapse in the memory of his intellectual operations, but not in their occurrence. It is the memory and not the conception of ideas that the poison has affected.

There should be two distinct faculties recognized in that which constitutes memory. For instance, one who is drunk enough to stagger home may remember the street, the house, and even the room where he lives. But the next morning he can remember nothing that occurred since he was drinking in the evening. How he got home is a mystery to him, and there is a void in his memory. Yet his memory was sufficient to enable him to find his way. There are, then, two kinds of memory, which we will call active and passive. The first is only possible when the intellectual faculties, including the will and attention, are intact. There must be power to direct the attention, and this power is absent in chloroform and alcoholic intoxication. Then, when poisons of the intelligence destroy the memory, they only alter the active, reflecting, conscious memory; their action is not exerted on the memory of past events (habit-memory), those are ineffaceable except by a profound lesion of the nervous centres. When chloroform passes into the blood through the pulmonary mucous membrane, active memory, which requires the will and attention, disappears, although the intelligence is not yet destroyed. Ideas are still conceived, old memories persist sometimes the recollection of past events is strangely sur-excited. Forgotten languages return, old events long buried in oblivion come back. This sur-excitation of the memory is a peculiarity sometimes met with in some forms of insanity where there is a coincident loss of active memory.

Chloroform-insensibility occurs very promptly, but it generally occurs only after the loss of the memory; and this loss of memory first, and sensation afterwards is very peculiar.

So when an operation is begun before insensibility is complete, the patient will shriek and struggle as though he suffered, and cry out that the operation has begun too soon. Yet, on awakening, he remembers nothing of what has occurred. The question arises—Is a pain that leaves no trace in the mind or memory a true pain ? It is difficult to answer, but we may assume that to suppress the prolonged consciousness of pain, is equivalent to suppressing the pain itself. A pain without memory is not a true pain, since it lacks that which is the precise characteristic of every painful impression, that prolonged sentiment which disturbs the mind, and the memory of which, whenever it occurs, is a feeble image, but nevertheless an active one, of the primitive pain. Two persons have their teeth extracted ; one with, the other without chloroform. Both appear to suffer alike; both cry out and struggle. But the one who has taken the chloroform remembers nothing of the pain, while the other, in imagination or memory, will suffer the operation over and over again.

When chloroform is given, a great deal depends on the disposition of the patient. If he is couragous and resolute, there will be no difficulty in producing insensibility ; but if he is filled with a dread of the operation, it is necessary to use great care : for we have noticed in such cases a greater tendency to syncope. Moreover, he will resist for a long time the action of the chloroform, so that more will be required than if he abandoned himself willingly to its effects. Although chloroform is irresistible, yet the cerebral excitation permits some patients to resist its toxic influence with a will that is exaggerated in strength. This same thing has its analogue in alcoholic intoxication. The more we study this agent the clearer it becomes that there is an antagonism between the different intellectual faculties, the voluntary and the involuntary faculties. The conception of ideas goes on as usual, but their normal correlation is lost. The association of ideas is continuous, and though the connection be disorderly and incongruous, still

there is no break or hiatus. Exterior sensations still occur
and bring their series of conceptions. Hearing is the last
sense to disappear. When the patient can neither see nor
feel, he hears what is said around him, and the remarks of
the bystanders give rise in his brain to all kinds of ideas.

(Something of the same kind occurs in the natural sleep
of children, rarely in adults. There is nearly always in
children a certain degree of natural somnambulism. The
child talks aloud, laughs with joy, or cries from fear. The
mother, by a few kind words and caresses, without waking
the child, can change the direction of these ideas, and quiet
the agitation so that the sleep becomes tranquil. On waking, there is no recollection of the occurrence.) But all
these phenomena which testify to the existence, if not the
integrity of the intelligence, do not delay its disappearance.

The groans, cries and laughter, are succeeded by confused and unintelligible speech. The muscles contract
with energy under the influence of delirium, they relax
slowly and end by becoming inert. To the period of
excitation succeeds the period during which there is profound sleep. Whatever may have been the violence of the
delirium or the severity of the operation which is performed, nothing can wake the patient from the comatose
state in which he is plunged. His respiration is regular,
the pulse is soft and full, the pupils are immobile, and the
features like one paralyzed. Intelligence is entirely abolished. Nevertheless, all parts of the cerebro-spinal nervous
system are not paralyzed. The integrity of the medulla
oblongata is indicated by the regular movements of the
heart and of the respiration, although the other parts of
the spinal cord are unable to accomplish their functions.
It is to this integrity of the medulla oblongata that immu-
nity from danger is due. It is therefore necessary to keep
a constant watch of the pulse and of the respiration, for too
strong a dose of chloroform may overwhelm that portion
of the nervous system which presides over the movements
of organic life. As to the spinal cord, it becomes affected

later than the brain, but sooner than the medulla oblongata, so that these three regions of the nervous system,
which preside over three different functions, appear to go
under the influence of chloroform singly and successively.
Claude Bernard has recently demonstrated that the brain
is paralyzed before the spinal cord, so that sensation is
abolished, while motility remains intact. He has shown,
moreover, that the brain may exercise over the spinal
cord a kind of paralyzing action. In limiting the action of
chloroform to the brain by a section of the spinal cord, we
obtain anæsthesia, but if we perform a reverse operation,
or, in other words, if we limit the action of chloroform to
the spinal cord, in preventing the encephalon from under-
going the action of the poison, anæsthesia will be impossible before the total death of the nervous cellules. (Confirmed also by the experiments of Lüys.) Very many other
volatile and toxic substances act in the same way as chloroform and may be used instead of it, but chloroform is the
type of all. The various forms of ether are most commonly
known and used. Certain gases have analogous properties,
and particularly nitrous oxide or laughing gas, which is
preferable to chloroform in dentistry and short operations,
because the anæsthesia occurs and passes off rapidly.

Recently a new substance, chloral, has been introduced
into therapeutics, which resembles chloroform in its chemical constitution and in some of its physiological properties.
It cannot take the place of chloroform in producing surgical anæsthesia, since it requires enormous doses to abolish
all trace of sensation, but it is a valuable hypnotic. In its
power to produce a tranquil sleep and to allay pain it
resembles morphine more than chloroform.

IV.

HASCHISCH.

The extract of Indian hemp, or haschisch, is used in three
forms: dawamesc, a nauseous mixture of aromatics, veget-

able oils and haschisch, taken before eating ; haschisch that is smoked in pipes or cigarettes, the true oriental style ; and the aqueous extract, or hafioun, which is more active than the other preparations. In moderate doses the effect is very agreeable and not dangerous. A slight disturbance of digestion, a little heaviness of the head, and cerebral excitation are all that is to be feared from ordinary doses of dawamesc and hafioun.

Unless anticipated, the first effects of haschisch may pass unperceived. These consist of a certain motor and sensitive excitation of the spinal cord. Twinges are felt in the neck, back, and limbs, and chills run all over the body. There are flashes hot and cold that rise to the head. With all this there is a certain comfort and satisfaction, like that accompanying a slight effect of alcohol. Little by little the excitation of the spinal cord produces more characteristic effects. They become uneasy, walk up and down, and stretch themselves out in all directions. They wish to dance, to move about, to lift enormous weights, and in the midst of this merely muscular excitement the mind remains calm. But all at once, by a chance word from one of the bystanders, by some perfectly natural remark, they are set to laughing. The laugh is involuntary, unreasonable, prolonged, nervous, convulsive, and seems interminable. When this burst of laughter has ceased, they know that it has been ridiculous : they regain their senses and understand that this laughter was one of the first effects of the poison.

From this moment the ideas become more and more crowded. It is like a perpetual blaze of fireworks, sparkling in all directions. One idea succeeds another with dazzling rapidity. Thoughts come and go in apparent disorder, but in reality according to the laws that govern the association of ideas and impressions. Their speech is agitated, almost furious : they are astonished to see that those around them do not partake of the same delirium which they feel, and are indignant at their stupidity.

They cannot express the thoughts that occur to them; language is not rapid enough to keep pace with their ideas.

Every idea, whatever its nature may be, is exaggerated, we may say that there is a hypertrophy of the ideas. We are moved to tears by what we should, in the normal state, consider merely annoying. The most simple things become theatrical, and it is with tragic tones that we announce that it is late, or that the wind blows. All this silly trash gives an infantile joy which we do not try to hide. We pass from laughter to tears without any transition period. Self-esteem is also exaggerated, and the man who has taken haschisch has become so superior to other men that he is filled with contempt at their stupidity.

There is also a complete moral transformation, and it is to be noted that all the phenomena resemble those of hysteria. In general, hysterical women have brilliant ideas, they are intelligent, the imagination is lively and fertile, but, however elevated their intelligence may be, it is defective for two principal reasons, the exaggeration of sentiment and the absence of the will. This double characteristic is found also in haschisch. In both, the exaggeration of sentiment produces immoderation, whether of joy or sorrow. Their self-conceit is such that the slightest remark, the most inoffensive word, wounds and offends them. They dramatize life. Simple, ordinary existence does not prevent them from indulging this theatrical tendency. They play with equal success either comedy or tragedy. Without the slightest occasion for display, they surprise every one by the suddenness, the mobility, and the intensity of their dramatic passion.

The impotence of the will is very remarkable in the hysterical. They have no control of themselves or their sentiments. They tell all that comes into their heads, giving utterance to a thought as soon as it is conceived, without regard to the consequences of their language. So, in a short conversation with one who is hysterical, we are

impressed with the contradictions, untruths, and extrava-
gances of thought, since the judgment and will do not in-
tervene to rectify that which is defective. That kind of
judgment-power or common-sense which enables us to
determine what is best to say and what is best to remain
unsaid, is unknown to the hysterical.

The same condition exists in the person who has taken
haschisch. It comes on suddenly, and therefore those who
have partaken of the poison should never allow themselves
to mix in company while under its effects, lest their extrava-
gant speech and actions should put them in some ridiculous
position that will not be understood by those around them.
Yet with the intoxication of haschisch there is a consciousness of the loss of will-power, and, unlike the condition of
those under the effects of alcohol, we are aware that we
cannot trust ourselves. There seems to be a double identity, one of ourselves is drunk with the poison, the other
looks on and sees and remembers it all. This consciousness of our condition is also found in some cases of insanity, for it is not uncommon for those unfortunates to come
to the asylum and request to be shut up lest they should
do themselves or some one else an injury, so conscious
are they of their condition.

The two most characteristic phenomena of haschisch intoxication and that which is not usually found in other
intoxications, is the alteration of our ideas of time and space.
Time appears to be immeasurably lengthened. Between
two clearly conceived ideas we think we entertain an infinite number of others, indistinct and incomplete, of which
we have a vague impression, but whose number and extent
fill us with admiration. These ideas seem innumerable,
and as time is only measured by the recollection of ideas,
time appears prodigiously long. As when one is awakened
suddenly by the fall of a canopy on the bed, the shock
gives rise to a series of fancies longer to recount than to
conceive of. The dreamer sees himself transported to the
top of a mountain surrounded by a hostile mob. They

throw him from the summit of a rock, and, after a descent
that seems to occupy years, he falls on his head in a ravine.
All these conceptions have occupied scarcely half a second,
the time that was necessary to be awakened by the piece
of wood that falls. So we may, by a kind of psychological
experiment, produce a similar illusion. If one while riding
in a carriage is overtaken by drowsiness which he tries to
resist, he will open and shut the eyelids at frequent intervals, and the distance passed over, as well as the time occupied while the eyes are shut, will seem enormously long.
There is no need of actual sleep to give rise to this illusion
as to the duration of time. While the eyes are closed, the
road we are on, or rather the time occupied in passing
over it, will seem interminable. Even one who is familiar
with the route, and knows that it is not very long, will
think he has already arrived, and each time he opens his
eyes there will be a new deception. In fact, when we
remain so self-concentrated, without seeing or hearing, we
only have a very imperfect notion of real time. On the
contrary, when our senses are awake and attentive, they
unceasingly correct the impressions founded solely on
psychic ideas. We only know very imperfectly the services which at each instant all our senses perform for us,
and it is only by reflection and psychological analysis that
we can hope to understand them. In dreams and sleep this
illusion as to the duration of time is vague, but with haschisch it becomes clearly marked. Just as astonishing is
the illusion which makes short distances appear immense.
It is not found in other kinds of intoxication, and I can
scarcely give a rational explanation of it. Even its description is difficult. The illusion makes a bridge or an avenue
appear to have no end, they stretch out to unheard of and
improbable distances. When we go up-stairs the steps
seem as high as the clouds. A river, whose opposite bank
is in plain sight, seems as large as an arm of the sea.

Besides these two illusions of space and time, which are
very tenacious and last often for more than twenty-four

hours, there are others equally strange, but perhaps not so invariable in their occurrence.

There is a remarkable resemblance between the illusions of haschisch and the systemic delirium of the insane. In the latter, the delirious idea has some real point of departure, a sensation or external impression. This is used as a basis for conceptions, and by a sort of induction, by no means illogical, a whole system of errors is built up. So, from the occurrence of nausea or gastric pain, they conclude that they have been poisoned and that we wish to kill them, and all the reasoning in the world would fail to convince them of their error. This is precisely what occurs in haschisch intoxication. The intelligence is entirely subordinate to the imagination and this, in turn, to an abnormal excitation of the senses. In our ordinary condition it is very certain that exterior impressions transform certain ideas and awaken others, but we are only conscious of just what we choose to be conscious of; the attention and the will eliminate and destroy outside impressions which we do not notice, so that they leave no trace in the intelligence. These two faculties cover with a thick veil all this unconscious work, and in the midst of the confused activity of intellectual operations the mind only sees what it wishes to see. Thus we can appreciate the condition of the intelligence in haschisch intoxication when the will and the attention no longer protect the mind from the inroad of every impression.

The feature which distinguishes intoxication of haschisch from that of alcohol is that the memory remains intact. We remember with astonishing exactitude all we have seen, said, and done. If the dose of the poison is very large however, there is a complete loss of memory and even furious delirium, so that haschisch has its dangers, although I do not know that any fatal case has been reported in Europe. In some cases the delirium lasts for many days and is very alarming. In the East, haschisch is in very general use. Almost always it is smoked in large pipes which are passed around. Its smoke is very agreeable, exhaling a peculiar aromatic odor. The effects of the smoke of haschisch do not differ essentially from those produced by the ingestion of the substance, but are commonly milder in their form.

V.

OPIUM.

The desiccated juice of the poppy, called opium, and its tincture, laudanum, have similar properties. We find however that opium is a very complex body, a mixture of many substances having analogous but not identical properties.

Besides narcotine and morphine, chemists have discovered many other constituents, such as codeine, narceine, thebaine, papaverine, etc., all of which enter into the composition of opium. But these different bases do not act upon the organic functions in the same way. Thus the soporific powers of narcotine are very slight, and two grammes of this substance may be taken without producing sensible results, although a centigramme of morphine, a dose two hundred times less in amount, is sufficient to provoke toxicological effects. Thebaine does not produce sleep, but excites in animals convulsions resembling those of strychnine; yet morphine in a similar dose brings on a profound sleep. Another remarkable action of the alkaloids of opium is the fact that their effects upon man are not the same as those they produce in animals. Man is peculiarly sensitive to the action of morphine, although thebaine produces hardly any effect on his nervous system. On the other hand, animals are not sensitive to the effects of even a very large dose of morphine, although thebaine is, for them, a virulent poison. Two grammes of morphine would not kill a dog, but ten centigrammes of thebaine would be inevitably fatal. In general physiology,

12

this difference of resistance to poisonous agents is not yet explained. We know that belladonna and atropine, so poisonous to man, have but a slight effect on a rabbit, and that a dose sufficient to kill ten robust men is hardly enough to kill a rabbit.

Morphine, the principal and most energetic substance contained in opium, so far as its action on man is concerned, has very nearly the same effect as crude opium. In medicine we prescribe them indifferently, so that we may comprehend them both in a common description.

The sleep-producing power of opium is supposed to be derived from its action on the cerebral circulation. It is not yet certain that this is the true cause, but it is the result of much laborious investigation.

On trephining the skull and inspecting the cerebral surface in animals, we find that the aspect varies according to the size and condition of the blood-vessels which cover the brain. Sometimes the surface is purple, swollen, and covered with a network of dilated blood-vessels, this is cerebral congestion. At other times it is pale, depressed, and the capillaries can hardly be distinguished, there is a deficiency of blood or anæmia of the brain. Now, it is found that the circulation of the retina of the eye is the image of the cerebral circulation, so that when the brain is congested, the eye is also congested, and *vice versa*. This furnishes a method of diagnosis by means of the ophthal-moscope, but there is another way of judging of the condition of the blood-vessels of the eye. The circular and contractile opening in the iris, the pupil, which contracts in the light and dilates in the dark, is always contracted when the brain is congested, and always dilated when the brain is anæmic, provided the observation is not made with the patient placed in a dazzling light, nor in a darkness too profound.

Then we might suppose that, since in normal sleep, as in the sleep produced by opium, the pupil is very much contracted, the brain is in a state of congestion in both cases,

and that sleep is the consequence of this cerebral congestion.

Unfortunately this theory is only an hypothesis, and
facts prove that it is not exact. Many physiologists, Durham and Hammond among others, have shown by experiments that during sleep there is anæmia of the brain.
According to them, we must understand that the afflux of
blood to an organ cannot determine a quiescent state of
the organ, and all physiological functions should be retarded by the diminution of the circulation, in the brain
as well as in all other vascular organs.

So we are unable, in spite of experiments, to arrive at a
definite conclusion as to whether opium produces congestion or anæmia of the brain.

Opium-sleep is not precisely the same as ordinary sleep.
About half an hour or an hour after having taken opium,
we feel a slight excitation, a general feeling of vivacity and
satisfaction, which is soon replaced by a kind of somno-
lence that is rather a condition of reverie than of dreams.
We experience a certain pleasure in permitting ourselves
to be overcome by the gentle torpor. Ideas become
images which succeed each other rapidly, without our
wishing to make an effort to change their course. So long
as the intoxication is not profound, this effort is still possible. We know that we are going to sleep, but that, if we
choose, we can overcome the drowsiness. Little by little
the legs grow heavy, the arms fall inert, and the heavy
eyelids cannot be kept open. We dream, we wander, and
yet we do not sleep; the consciousness of the exterior
world has not yet disappeared. The noises outside, the
tick of the clock, the rumbling of carriages, are obscurely
perceived. Active, conscious individuality does not exist;
a passive condition has replaced it. Gradually all becomes
more vague, the ideas are lost in a confused mist, and
everything seems immaterial. The body seems to have
vanished, and only thought remains, brilliant, but growing
more and more confused. Then all exterior impressions

disappear, and we experience either febrile agitation, delirium, or, more commonly, a calm and tranquil sleep. The charm of this condition is that we know we are sleeping. The sleep is intelligent and comprehends itself. The hours pass with marvellous rapidity, and in the morning when opium seems to have spent its force, but in reality preserves it in full, the sleep is peculiarly delightful. The mind, freed from all terrestrial bonds, seems to reign in a world of tranquil and serene thought. This intoxication is wholly psychic, far superior to that of alcohol or haschisch, for while hashish gives a few hours of lunacy, opium gives sleep, which is far better.

One should have suffered from insomnia in order to appreciate what opium can do. To listen while all the minutes of the night successively pass in the midst of an overwhelming silence, to turn and toss on the bed, to sketch the outline of confused ideas, without being able to grasp a single one, to struggle hopelessly against an unconquerable restlessness, this is torture which no one can comprehend who has not experienced it. The power which opium possesses to give a calm and tranquil sleep, to assuage pain, and to control the febrile agitation in some diseases, renders it a priceless boon to the patient and a remedy of wonderful value to the physician. It is not merely a poison of the intelligence, it exercises also an energetic modifying action on the sensibility. We do not know whether this comes from its action on the nerve which transmits the excitation, or on the brain which perceives it, but without causing sleep it still has the power of calming nervous excitation and hyperæsthesia. When it quiets hyperæsthesia it does not procure sleep, for all its force seems to be exerted against the pain, and there is not enough action to carry its effect so far as to induce sleep. Those who take opium to allay inveterate neuralgia, are obliged to increase the dose to a great extent before they can obtain its narcotic action.

There is in one respect a strong contrast between opium

and alcohol, for alcohol has a cumulative effect; the more one drinks the more easily they become intoxicated, but with opium the system will become accustomed to the poison, so that increased doses will be necessary to produce the effects at first obtained by moderate doses. When one has become habituated to the use of opium, it becomes a necessary stimulant, and he becomes just as sick from an absence of the poison as from its excess. So if by chance we diminish the dose or forget to give it, such patients are seized with sudden attacks of pain, whose origin is easily traced to the absence of the stimulant to which the system has been accustomed.

Opium is a dangerous poison, not merely from the fatal effects of an overdose, but because its use is apt to lead to the opium habit. In China this is a national vice; and the opium-shops, where opium is smoked, correspond to the cabarets and taverns of more civilized lands. Travellers tell us that opium-smokers who daily indulge in this excess fall at last to the lowest depths of degradation, moral and physical. Pale, haggard, and emaciated, scarcely able to drag themselves around, they can only arouse a little energy when a new dose of the poison stimulates them.

VI.

COFFEE.

Opium has an antidote, so that while we are able produce sleep, we are also able to produce insomnia. The effects of coffee on the intellect are diametrically opposed to those of opium. Its action is well known, because it is in common use. Some use it as a necessary stimulant to intellectual work. In others this excitation is manifested by a painful insomnia. For them, coffee is a veritable poison that deprives them of the most precious of boons. In some cases, when they have taken too strong a dose, a most distressing condition of anxiety and restlessness comes on, wholly different from the quiet activity of opium. With coffee the imagination is scarcely excited, but the will seems to be stimulated. We wish to walk quickly, we cannot keep quiet long enough to finish any task. With opium, alcohol, and haschisch we feel that there is a kind of drowsiness of the attention, but with coffee the attention and memory are perpetually on the alert. It gives, therefore, a veritable intoxication which is much more fatiguing than the somnolent intoxication of opium, but it leads to the same result. In wishing to do too much, the mind does too little. By this over-excitement the will is injured, and the perfect balance of the intellectual faculties is destroyed as much by excess as by inactivity of the will.

We say generally that coffee produces anæmia of the brain, while opium and alcohol cause congestion, but this physiological theory is far from being established on a sure foundation, and new experiments are necessary. Nevertheless, we know very exactly the *rôle* played by coffee in general nutrition. It delays organic combustion and economizes food. In the normal state an infinite number of chemical actions are carried on in the system whose final results are the production of heat and the liberation of carbonic acid. The carbonic acid passes into the venous blood from which it is eliminated by the lungs. The quantity of carbonic acid, then, may in a measure be regarded as an indication of the activity of nutrition. Now with coffee, we find that the quantity of carbonic acid is diminished without any diminution of the vital forces, and this occurs without any change in food, work, or functional activity. Instances might be given of the use often made of coffee, to enable workmen to continue laborious tasks with little food. It therefore moderates the activity of the chemical processes and prevents the waste of tissue. Other substances have analogous properties, especially tea and cocoa. It is probable that caffeine, theine, and cocaine are analogous chemically and physiologically, and that their effects on the intellectual functions are very nearly identical.